The Scholarship Series in Biology

General Editor: W. H. Dowdeswell

Ecology of Fresh Water

THE SCHOLARSHIP SERIES IN BIOLOGY

Human Blood Groups and Inheritance
S. Lawler and L. J. Lawler

Woodland Ecology
E. G. Neal

New Concepts in Flowering Plant Taxonomy
J. Heslop Harrison

The Mechanism of Evolution
W. H. Dowdeswell

Heathland Ecology
C. P. Friedlander

The Organization of the Central Nervous System
C. V. Brewer

Plant Metabolism
G. A. Strafford

Comparative Physiology of Vertebrate Respiration
G. M. Hughes

Animal Body Fluids and their Regulation
A. P. M. Lockwood

Life on the Sea-shore
A. J. Southward

Chalkland Ecology
John Sankey

Ecology of Parasites
N. A. Croll

Physiological Aspects of Water and Plant Life
W. M. M. Baron

Biological Principles in Fermentation
J. G. Carr

Studies in Invertebrate Behaviour
S. M. Evans

Ecology of Refuse Tips
Arnold Darlington

Plant Growth
Michael Black and Jack Edelman

Animal Taxonomy
T. H. Savory

Ecology of Fresh Water
Alison Leadley Brown

Ecology of Fresh Water

Alison Leadley Brown, M.A., F.I.Biol.

H·E·B

1971

Heinemann Educational Books Ltd
London

Heinemann Educational Books Ltd

LONDON EDINBURGH MELBOURNE TORONTO
AUCKLAND JOHANNESBURG SINGAPORE
HONG KONG IBADAN NAIROBI NEW DELHI

ISBN 0 435 61537 8

Published by Heinemann Educational Books Ltd
48 Charles Street, London W1X 8AH
Printed in Great Britain by
Richard Clay (The Chaucer Press) Ltd, Bungay, Suffolk

Preface

The word 'ecology' has become common usage, applied not only in the biological context but also to many other subjects, and the fact that it has been used in the title of this book, perhaps needs some explanation.

By definition, ecology means the study of animals and plants in relation to their environment. Two, or perhaps three, decades ago this was a relatively new approach, since books which had been published prior to this period, relied upon a straightforward account of structure and function without reference to how they fitted an organism to exist in its chosen habitat.

An ecological study roams over many branches of science and yet must inevitably be confined by its own principles. The purpose of this book is to show how these principles can be applied to the study of different kinds of fresh water.

Although the various factors affecting the distribution of freshwater animals and plants are discussed in theory, the author has endeavoured to illustrate such theoretical considerations by examples of field-work actually carried out in still and flowing water. Also introduced are investigations of some of the many microhabitats which offer enormous possibilities for original work.

No amount of reading can be a substitute for actual experience in the field and if encouragement to students has been given to abandon the library and find out for themselves, then one of the main objects of this book has been achieved. Indeed, it is largely through the endeavours of students from Rolle College on the various expeditions we made together to the rivers, streams, and ponds of Devon and Somerset, that this book has been written. To them thanks are due for, maybe unwittingly, they have contributed much of the practical material upon which the book is based.

I am grateful also to Mr. W. H. Dowdeswell, the editor of the series, for his constructive criticism of the manuscript and for the encouragement he has given.

With the exception of plate 2(*b*), for which thanks are due to John Clegg, all other photographs are by the author.

1971   A.L.B.

Contents

PREFACE   v

LIST OF PLATES   viii

INTRODUCTION   ix

1 The environment of fresh water   1

2 Aquatic plants and the environment of fresh water   12

3 The animal inhabitants of a pond   25

4 Factors affecting the distribution of freshwater organisms   42

5 Studies of spatial distribution and density
     (1) In still water   57

6 Studies of spatial distribution and density
     (2) In running water   67

7 Trophic relationships   88

8 The flow of energy through an ecosystem   97

9 The influence of man on freshwater communities   103

APPENDIX I   Table of solubility of oxygen in chloride-free water at various temperatures   113

APPENDIX II   Methods of measuring dissolved oxygen   114
                            (a) Winkler method
                            (b) Protech portable dissolved oxygen and temperature meter

APPENDIX III Measurement of flow by means of a current-meter   117

BOOKS FOR FURTHER READING   119

REFERENCES   120

INDEX   123

List of Plates

opposite page

Plate 1   WESTERN BLADDERWORT (*Utricularia neglecta*)   20
(*a*) Submerged frond
(*b*) Portion of frond

Plate 2   WATER SOLDIER (*Stratiotes aloides*)   21
(*a*) Plants in early summer
(*b*) Plants in late summer

Plate 3   BRACKISH DYKES AT TOPSHAM, DEVON, IN JULY   52
(*a*) A dyke near its outflow into the estuary
(*b*) A dyke in the upper water meadows

Plate 4   GREAT SILVER WATER BEETLE (*Hydrophilus piceus*)   53
(*a*) Egg cocoon
(*b*) Adult beetle

Plate 5   THE HAUNT OF THE GREAT SILVER WATER BEETLE   84
(*Hydrophilus piceus*)
(*a*) Water violet (*Hottonia palustris*)
(*b*) Well-vegetated peaty rhine

Plate 6   SLOW-FLOWING WATER   85
(*a*) The Tiverton Canal, Devon, in June
(*b*) A calm reach of the River Culm, Devon, in June

Plate 7   EXTON STREAM, DEVON   100
(*a*) Weir and pool
(*b*) Deep and shallow reach

Plate 8   STREAM AND RIVER CURRENTS   101
(*a*) A swift reach of the River Otter, Devon
(*b*) Using a current-meter to record rate of flow

Introduction

Lord Balfour once said 'The great bulk of people infinitely prefer the continuance of a problem which they cannot explain to an explanation which they cannot understand.' The study of fresh-water habitats certainly poses many problems which have not yet been explained and many whose explanation is difficult to understand. Nevertheless, to the biologist there is a wealth of interest in the world of fresh water which can lead him to innumerable investigations.

Ecology can be simply defined as the study of organisms in relation to their environment, yet only during the last few decades have biologists concerned themselves with the question of the interaction of living organisms with their surroundings.

The population of any stable habitat is composed of plants, herbivores, carnivores, and parasites. If there were only plants and hervibores present, the animals would increase by devouring the plants until they starved, leaving a few survivors to start the cycle once again. Such an unstable state of affairs cannot be repeated indefinitely. It is the way in which organisms interact to produce stability within a habitat that is the main interest of the ecologist.

Life for plants and animals in any medium is a constant struggle. Besides the physical environment, predators, parasites, and other competitors have to be contended with; but compared with a terrestrial existence, fresh water offers a relatively stable environment.

It is generally accepted that life began in the sea and that the colonization of the land masses took place by invasion from the sea. It might seem even more probable that colonization of fresh water was also directly from the sea.

The sea offers about the most constant medium in which to live, while fresh water is liable to greater variation of both temperature and salt concentration, especially after heavy rain. Moreover conditions may vary from one stretch of fresh water to another.

Such physical factors have in the past proved an effective barrier to all but a few organisms in broaching fresh water from the sea. Some species of fish, a few molluscs, and the Freshwater shrimp (*Mysis relicta*) are examples of those which have been successful. From an examination of the different animal phyla to which the present-day species of freshwater animals belong, it is evident that invasion of fresh water has largely been from the land and for all freshwater plants this is the case. There are also one or two organisms that live in the sea but which migrate to fresh water, or vice versa, in order to breed. The best known are the Common eel (*Anguilla anguilla*), which breeds in the sea, and the Salmon (*Salmo salar*), which breeds in fresh water.

Invaders of fresh water have been presented with a number of problems and no single species has solved them all. The degree of adaptation imposed by their new environment varies from practically no alteration in structure, which is the case in some of the water beetles, to considerable modification of both anatomy and physiology needed by species inhabiting the swift current of an upland stream or by those living in mud where there may be a complete absence of oxygen.

Plants and animals do not live in isolation but in breeding populations. More often several populations, composed of different species, will be found living together as a community. The underwater parts of a plant such as the Water parsnip (*Berula erecta*), with its branching roots, can support populations of the Water louse (*Asellus aquaticus*), the Freshwater shrimp (*Gammarus pulex*), and many other organisms including algae upon which the animal populations feed. Such a community of animals and plants is occupying a habitat—in this case the roots of *B. erecta*. However, the term 'habitat' is often used in a variety of contexts and can embrace many features besides the actual living space of the particular animal or plant. It is also true to say that the bound-aries of a habitat are often difficult to define.

From a study of different habitats it becomes clear that each supports different populations. Not only can a community of organisms be associated with a particular habitat, but the habitat can be modified by the activities of the organisms themselves. In this way the community and the habitat evolve together to form what is called an *ecosystem* within which there is an economic association of its living components. Words such as 'community' and 'population' are often used rather loosely in describing an ecosystem. This confusion can become greater when considering the composition of a microhabitat. For instance, a hole in a tree stump temporarily filled by rain-water can support a highly typical fauna and flora. In fact the uniformity of conditions characteristically prevailing within such a microhabitat means, in turn, a restriction in the number of species adapted to live there and it is open to question whether the groups of organisms should be described as communities. The term ecosystem, as applied to freshwater habitats will be more easily understood as these are described.

It must not be forgotten that the ecology of animals is inextric-ably interwoven with their behaviour, so that in many ways their distribution is a record of their behavioural reactions. Whereas the distribution of plants is governed largely by their physical environment, that of animals is a much more selective process. This often involves us in a study of the animal's behaviour in order to interpret the reasons for its distribution.

Finally, it must be remembered that the ecological study of an environment may involve many other branches of science, roam-ing over the preserves of the physicist, chemist, physiologist, and taxonomist. Yet ecology is a science in its own right carrying its own disciplines and principles. The aims of this book are to show how these principles can be applied to the study of fresh water.

1

The environment of fresh water

The environment of organisms living in fresh water consists of a number of habitat factors such as the temperature, the amount of sunlight penetrating the water, its density and its chemical constitution all of which influence their distribution on, and which interact to cause fluctuations in, the size of populations.

Because they never act on their own, it is often difficult to separate the effect of one factor from that of another and it is easy to attribute the adaptations made by an organism to a particular factor when they are probably due to the interaction of several.

For instance, the current in a fast-flowing stream may seem to be the controlling factor causing certain organisms such as the nymphs of the mayfly *Ecdyonurus*, to seek the shelter of the underside of a stone. Further investigations may well show that *Ecdyonurus* orientates itself also to the amount of light falling on it and during the hours of daylight it is usually to be found beneath the stones in the stream bed, moving round to the top of the stones during darkness. Its habits and its flattened body and limbs are the adaptations made in response to a combination of influences, no one factor being solely responsible. Nevertheless, we can rightly assume that some factors have a greater influence than others on the organisms living in a particular habitat.

In a brief survey of this kind there is not room to make an elaborate study of the whole complex of factors affecting freshwater animals and plants. We must single out those which play an important part in bringing about the distribution of organisms or which control their numbers.

Chemical factors

The origin of all the fresh water of ponds, lakes, and streams is the atmosphere in the form of rain. Apart from a small amount of dissolved carbon dioxide, rain-water is, however, devoid of most substances essential for the maintenance of life. Fortunately very little of the water reaching these bodies of water actually falls directly on them as rain. Most of the water arrives as the result of drainage from the surrounding land and as it travels towards a lake or pond, gaseous and solid substances become dissolved in it, enriching the water and making it suitable as a medium which can be colonized by animals and plants.

The composition of fresh water varies with each body of water and we can only mention some of the substances which play an important part in the lives of animals and plants living in it.

Oxygen

To all but a few organisms capable of living in anaerobic conditions, oxygen is essential to their life processes. The gas becomes dissolved at the surface where air is in contact with the water. The photosynthetic activities of water plants also augment the supply. On a sunny day in a pond containing a number of water plants, the water can become super-saturated with oxygen in the region of the weeds. At night, when photosynthesis ceases, the dissolved oxygen is depleted by the respiratory processes of the animals and of the plants themselves.

The amount of oxygen that a given volume of water will hold in equilibrium decreases as the temperature increases, and the concentration of oxygen in water of a certain temperature is usually expressed as the percentage of what it would be if the water was saturated with air at normal pressure. A table of solubility of oxygen at different temperatures is given in Appendix I.

From the graph in figure 1 it is evident that even when water is saturated with oxygen it contains very little of the gas. For instance, at $5°C$ 1 l of water contains only 9 cm^3 of oxygen while at $22°C$, 1 l contains 6 cm^3. Air, however, is a much richer source of oxygen, 1 l containing about 210 cm^3. With a rise of tempera-

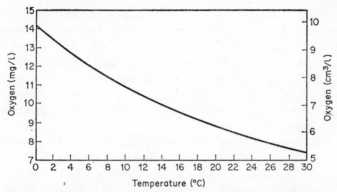

Fig. 1. The oxygen content of water saturated with air at normal pressure (760 mmHg). (From Mortimer, 1956.)

ture the rate of metabolism of many invertebrates may be trebled and an oxygen shortage can easily be created in the water.

A shortage of oxygen can prevail in water which contains a large amount of decaying animal and plant matter, for the process of decomposition uses up the oxygen present. In a stagnant pool oxygen deficiency may be so great that it cannot support any life except for those forms such as certain aquatic worms and the larvae of a few species of insects, which possess special adaptations for living in such conditions.

Some aquatic insect larvae and other small animals which have a thin body covering can make a direct exchange of oxygen between their body tissues and the surrounding water. Others have developed gills of various kinds by means of which they can obtain dissolved oxygen from the water. These modifications are discussed in Chapter 4.

Mineral salts

The nature and the amount of mineral salts present in any body of fresh water will depend upon the geology of the surrounding land since they enter in the drainage water from the soil. Mineral nutrients are also regenerated from the bottom mud by bacterial action.

Carbon dioxide is very soluble in water and can enter direct from the atmosphere. Other sources are the bacterial decomposition of organic substances within the water or the result of the respiration of aquatic organisms. Carbon dioxide combines readily with the salts of calcium and magnesium to be 'stored' in the water temporarily as soluble bicarbonates or more permanently as carbonates which sink to the bottom to become mixed with the mud.

The amount of bicarbonate and carbonate present is a measure of the total alkalinity of the water and is limited, not by the amount of carbon dioxide which is always available, but by the supply of calcium and magnesium.

In limestone districts the soil will contain considerable quantities of calcium salts which, in the presence of carbon dioxide, become changed into bicarbonates and carbonates. Water with a high concentration of these salts is alkaline and is called 'hard' water. A chalk stream may contain as much as 100 mg/l of calcium. Some creatures prefer hard water while others do not. Freshwater sponges, most crustaceans, and many freshwater snails, whose shells are largely composed of calcium carbonate, thrive in hard water. On the whole such water will support a wider variety of animal life than the calcium-deficient water of moorland bogs and streams which originate from acid, peaty soils.

Hydrogen ion concentration

Water dissociates according to the following equation:

$$H_2O \rightleftharpoons H^+ \times OH^-$$

and the number of grammes of hydrogen ions present per litre is a measure of the acidity or alkalinity (pH) of that solution. Hard water with a high calcium content has a high pH value, while 'soft' water has a low pH.

During daylight when the plants are actively photosynthesizing, the pH of pond water can be greatly increased by the removal of carbon dioxide. Conversely, the production, as a result of decomposition of large quantities of carbon dioxide and other organic acids at the bottom of a pond will cause the water in this region to be acid with a low pH. Peaty moorland pools can have a pH of 4·6 or even lower. The pH of flowing water tends to remain fairly constant while in ponds and lakes, for the reasons outlined above, and because of seasonal changes in the amount of weed, there will be fluctuating conditions.

Seasonal fluctuation in chemical conditions

Between winter and summer there can be great chemical changes occurring in bodies of fresh water. Fluctuations in the temperature of the water involve, as we have already seen, changes in the amount of dissolved oxygen present. Likewise there can be seasonal changes in the amount of dissolved carbon dioxide.

During the winter when the photosynthetic activities of the plants are at a minimum, the total alkalinity of a pond or lake can increase enormously as the amount of stored carbonate increases. In summer, due to increased photosynthesis, there is a depletion in the amount of carbon dioxide and carbonates are precipitated as the temperature rises, further reducing the total alkalinity. This in turn means an increase in pH. Figure 2 shows the seasonal variations in the amount of dissolved oxygen, pH, total alkalinity, and carbon dioxide acidity (the concentration of carbon dioxide in solution as carbonic acid), in a shallow lake over a period of four years. From this graph it can be seen that there is a definite correlation between the four factors, as well as winter and summer fluctuations throughout the period.

Such chemical changes which inevitably occur in any kind of fresh water, have a profound effect upon the organisms and upon their food relationships.

Physical factors
Temperature

Since water is heaviest at 4°C and becomes increasingly lighter until it freezes, ice will form on the surface of a pond but will rarely extend very far down. Because of this, most freshwater

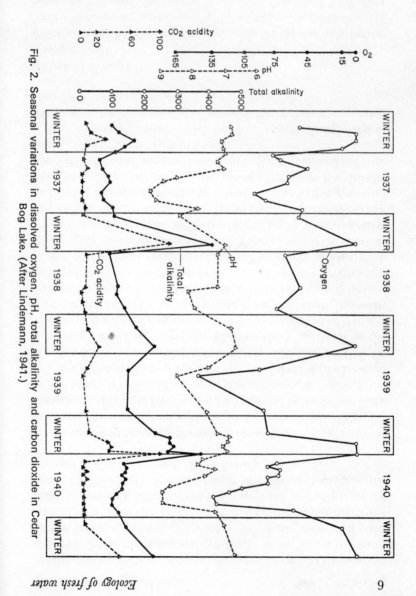

Fig. 2. Seasonal variations in dissolved oxygen, pH, total alkalinity and carbon dioxide in Cedar Bog Lake. (After Lindemann, 1941.)

animals, unlike their terrestrial relatives, never experience a temperature below freezing-point.

The sun is the source of heat by which most water is warmed. In small bodies of water, that at the surface becomes warmer and this will soon be mixed, due to wind turbulence as well as to convection currents, with the water below. A point may be reached during a summer day when there will be only a very small temperature gradient between surface and bottom, but during the night heat will soon be lost from the upper layers by radiation. There may, therefore, be a considerable diurnal fluctuation in temperature which will follow the air temperature fairly closely with only a small lag in amplitude and time.

The relationship between air and water temperatures in a small body of water, at its deepest point 3 m, is shown in figure 3. Measurements of the water temperature were made at a depth of 15–20 cm below the surface and the air and water temperatures were recorded from March to August. Comparison of average water temperature with the weekly average of the air temperature, based on the daily maxima and minima, shows that the two run fairly parallel although daily records would probably reveal a rise in one while the other falls. For the months of May to August the water was warmer than the air. [For methods of measuring lake temperatures by means of a thermistor, the paper by Mortimer and More should be consulted.*]

Lakes with a large volume of water and a greater depth, present a different picture. Following the coldest part of the winter there may be a period when the lake will have a uniform temperature of 4°C from top to bottom. As the air temperature rises in the spring, the upper layers will be warmed but since the sun's rays are soon absorbed by water, the deeper parts remain cold. A stormy period may cause the warm upper and cold lower waters to mix, but as the summer progresses the two layers become established, with such a big temperature difference between them that they remain separate for the rest of the warm period of the

*Mortimer, C. H., and Moore W. K. (1953). 'The use of thermisters for the measurement of lake temperatures' *Ass. Theoret. and Applied Limnol.*, No.2

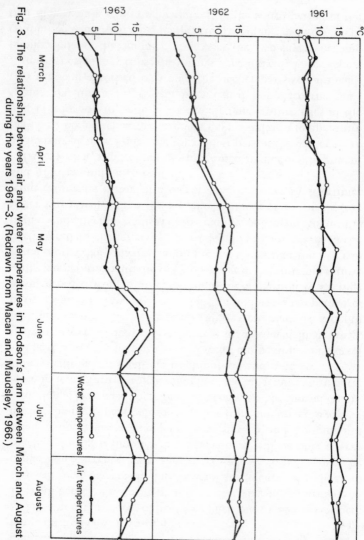

Fig. 3. The relationship between air and water temperatures in Hodson's Tarn between March and August during the years 1961–3. (Redrawn from Macan and Maudsley, 1966.)

year. These conditions can prevail in a large lake such as Windermere. Figure 4 shows that by midsummer the temperature of the lower zone of the lake, or *hypolimnion*, varies only about 2°C between a depth of 15 and 60 m and that that of the upper zone, or *epilimnion*, only 1°C from the surface down to about 8 m. In between there is a zone of rapid change called the *thermocline*, in which the temperature falls rapidly. During the cold nights of autumn, heat is soon lost from the surface and the epilimnion cools down and is eventually completely mixed with the hypolimnion.

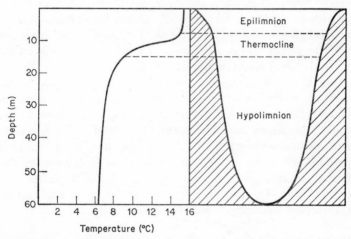

Fig. 4. Temperatures taken on a July day in Lake Windermere at different depths. (Redrawn from Macan and Worthington, 1951.)

Light

Rays of light falling on the surface of water do not penetrate very far, and sedimentary matter, even the organisms themselves, will absorb light.

Since the photosynthetic activities of plants are dependent upon light, plant growth is only possible where they can receive it. Although in Lake Windermere light penetrates sufficiently far to allow the growth of phytoplankton to a depth of 10 m, in a pond the amount penetrating a thick bed of rooted and floating weed will decrease rapidly from surface to bottom. This, in turn, will mean that while those plants at the surface will quickly reach a maximum rate of photosynthesis during the hours of daylight, those near the bottom, because the dense weed occludes the rays, will have a much lower rate of photosynthesis and hence a longer compensation period (the time from first light each day that the plants take to regain the carbon dioxide lost by respiration each night).

The amount of oxygen produced by pond weeds during photosynthesis can be used to measure its rate. Oxygen concentrations can be measured either by the Winkler method (Appendix II(a)) or with an oxygen probe (Appendix II(b)). Using either of these methods, it can be shown that the production of oxygen is greater when there is a high concentration of weed near the surface than when the weed is evenly distributed at all depths.

Light, therefore, directly controls the growth of plants and since plant life forms the basis of animal foods chains, a stretch of water rich in plant life will support a large animal community. Light also enables animals to see and to be seen, which is important when considering food relationships and predation. It is perhaps also worth noting that neither those freshwater invertebrates which possess compound eyes nor any of the fishes have colour vision. This fact may account for the rather dull and inconspicuous colouring of most freshwater animals.

Water movement

In the static water of a pond or lake, movement of the water is brought about by wind and by convection currents. This causes, to varying degrees, the mixing of the upper and lower layers. More violent wind action may result in waves which erode the banks depositing silt and mud.

Current in a stream or river results in the constant erosion of the stream bed, sand and silt being washed away to settle in the neighbourhood of objects such as large stones or where the current is slacker.

The current exerts a profound influence on the distribution of organisms inhabiting flowing water. Rooted plants and sessile animals, except in extreme cases of flooding, manage to remain in one place. For the rest, it is a question of possessing special structures for clinging to stones or other objects which prevent them from being washed away. For these reasons it is true to say that the number of species which have been successful in modifying their behaviour and structure to living in flowing water are far fewer than those which are to be found in static water.

By the constant churning action of the current, the water of a stream can be almost permanently saturated with oxygen. The temperature, the mineral content, and the pH all remain relatively uniform, a state of affairs which never prevails in the static water of a pond or lake.

Successful freshwater organisms must be able to withstand the hazards, as well as exploit the benefits, of their habitat. The ways in which they do so are the subject of the two succeeding chapters.

2

Aquatic plants and the environment of fresh water

The plants associated with inland waters include species from all the plant phyla. Microscopic algae, such as diatoms and desmids, form the phytoplankton, besides the numerous species of macroscopic algae, many of which live epiphytically upon other plants. A few liverworts live on the damp banks of ponds and ditches and one species, *Riccia fluitans*, is occasionally found in association with floating duckweeds. Several small species of moss grow on rocks or stones in streams but the true water mosses, belonging to the genus *Fontinalis*, are often quite large. The Water fern (*Azolla filiculoides*) introduced from the United States, has now become naturalized in a few places in this country. The horsetails, most of which are typical marsh plants, have several truly aquatic species, while there are a large number of angiosperms inhabiting marshland, many which are true aquatics.

Water plants are of fundamental importance to the animal life, for not only do they serve as a source of organic food but as a result of their photosynthetic activities, they give off oxygen required by animals for their respiration. If their leaves and other green parts are submerged, the oxygen dissolves in the surrounding water becoming available to those animals which, by means of gills or other respiratory organs, are able to absorb the dissolved oxygen. Aquatic plants also offer shelter to small animals and their underwater stems, leaves, and roots provide places on which animals can deposit their eggs.

The greatest advantage bestowed on plants living in or near water is that there can be little danger of drought or of transpiration exceeding water intake, always a serious risk to most terrestrial plants.

Nutrient salts

Besides water, plants require carbon dioxide and light in order to manufacture their food, combining the carbohydrates so formed with other inorganic substances.

Aquatic plants depend directly upon the same inorganic salts as those used by land plants, but these are present in solution in the water and are therefore readily available. The raw materials that they take from it include nitrates, phosphates, sulphates, carbonates, silicates, and traces of other inorganic salts. It was mentioned in the last chapter that the chief source of these inorganic compounds is via drainage from the surrounding land as well as by the decomposition of living matter and from the organic waste excreted by the organisms themselves.

The term 'nutrient' is often employed rather vaguely to describe substances which are used by plants in different ways. A substance may be employed as a source of energy or as a source of elements such as carbon, nitrogen, phosphorus and so forth. It may also supply small quantities of a particular chemical group of the vitamin or growth factor type.

Nevertheless, whatever the origin of these plant nutrients, their concentration varies not only between one body of water and another or even between different parts of the same stretch of water, but also with the season of the year. These fluctuations have a direct influence on the distribution and abundance of aquatic plants which are, themselves, largely responsible for these changes.

The quantities of most inorganic salts in fresh waters are very low and apart from nitrates, which are usually present in higher concentrations, can only be estimated in parts per million. Nitrates are used by aquatic plants for the formation of proteins so that when rapid growth begins in the spring much of the nitrate content of the water is removed to be restored in the autumn and winter by bacterial action in the breakdown of decaying plant and animal material. Concentrations of phosphates in solution follow much the same fluctuations as those of nitrates and may remain at practically nil all the summer.

Seasonal variation of the silicates is interesting for it corresponds to the periods of maximum development of diatoms in the phytoplankton such as *Asterionella formosa*, which has a skeleton composed of silica. Silicates are removed by *Asterionella*, the most abundant of the silicious diatoms, from solution in the water and as their numbers increase, a point is reached when silicates are removed more quickly than they can be replaced by the water entering a lake. Figure 5 shows the rapid decrease in silicates during the early summer, corresponding to the rapid increase in

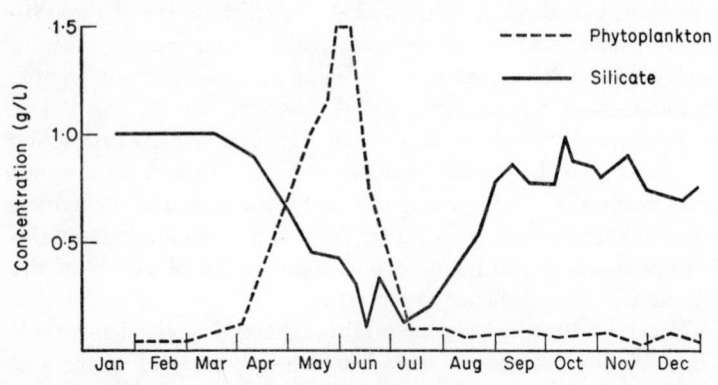

Fig. 5. Fluctuation in phytoplankton with concentration of silicates in Lake Windermere during 1936. (After Macan and Worthington, 1951.)

phytoplankton. When the silicates reach a critical level the numbers of phytoplankton fall steeply, due to the inability of *Asterionella* to extract silicate in low concentration. In the autumn there may be an increase in the silicates which could be attributable to a rise in planktonic crustaceans which feed upon *Asterionella* rejecting the skeleton.

The bladderworts *Utricularia* spp. float, totally submerged just beneath the surface and have no true roots. (See Plate 1.) They frequent water such as peaty pools which are rather poor in mineral salts and make good this deficiency in another way. Small bladders are borne on the hair-like leaves. Each bladder has a circular aperture at the end surrounded by a funnel of bristles and closed by a trap-door opening inwards. Hairs lining the inside of the bladder abstract the cell sap and when the trap-door is closed this causes a tension within the bladder which contracts. Small crustaceans, attracted it is thought by mucilage, touch the sensitive bristles around the opening. This causes the release of the trap-door and the rapid filling of the bladder with water which draws in the small organisms. After death within the bladder their decomposed remains are absorbed by the plant, in this way compensating for the lack of minerals in the surrounding water, particularly nitrates and phosphates, the chief constituents of protein.

Calcium carbonate also diminishes very markedly during the season of greatest plant activity. The plants abstract carbon dioxide and the soluble bicarbonate from the water leaving the insoluble acid carbonate as a precipitate. In waters with a high calcium content this precipitate can sometimes be noticed as a scaly coating on the leaves of water plants such as the Water soldier (*Stratiotes aloides*). (See Plate 2.) The plants float just beneath the surface for most of the year, their long roots hanging down into the water to maintain the plant in a level position. If they are moved to another pond, they will heel over until they regain their level position. In early summer, young leaves are produced which increase the air-space/tissue-volume ratio and hence their buoyancy, so that the plants rise to bring the new leaves above the surface. White flowers appear at this time, but in Britain they are all females and no seed is set. Normally the plants reproduce vegetatively by growing plantlets on the end of stolons.

Many plants can accumulate certain substances which they absorb from the water. Such a process takes place in the Stonewort (*Nitella clavata*). If it is placed in water containing 26 mg/l calcium ions, the cell sap is found to contain 380 mg/l of the ions. The ability of the stoneworts to accumulate salts of calcium accounts for their name and for the fact that their whorls of leaves are often covered with a heavy deposit of calcium carbonate.

The fluctuation in the concentration of salts is greatly influenced by plant activity which in the end reacts on the plants themselves, for depletion of the salts checks multiplication until decay once more restores the balance. The seasonal activity of the plants and variation in the concentration of salts is a reciprocal process. The salts reach their highest level at a time of the year when plant growth is at its lowest ebb and decrease in amount with increased uptake by the plants when renewed growth begins.

Temperature

The amount of life which any body of water can support ultimately depends on the supply of inorganic salts, but temperature certainly plays a part in the ability of the plants to absorb minerals such as potassium. The graph in figure 6 shows the effect of temperature on the uptake of potassium by young barley roots, the rate of uptake increasing steeply to 30°C and then starting to fall off. This is also typical of a number of water plants including the stoneworts and the Canadian pondweed (*Elodea canadensis*).

As we have already seen temperature also affects the amount of oxygen which the water can contain in solution. Like animals plants require oxygen for respiratory purposes deriving therefrom energy for growth.

Light

Since light is necessary for photosynthesis, the amount reaching the plants growing at different depths is part of the complex of factors influencing the degree to which the raw materials can be utilized.

Mention was made in Chapter 1 of the effect of light intensity on the rate of photosynthesis. Plants living in areas of high light intensity, either at the surface or in regions beneath the surface where the weed cover is scant, will have a high rate of photosynthesis and therefore a short compensation period, or a shorter time to recover the carbon dioxide lost through respiration during the night. On the other hand, those plants living in regions of low light intensity because of the slower rate of photosynthesis, will take longer to regain the carbon dioxide lost during the night

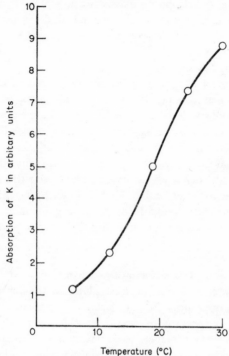

Fig. 6. Temperature curve for the absorption of potassium by young excised barley roots. (From Hoagland, 1948.)

and therefore have a longer compensation period which is relative to the degree of light intensity.

At the surface of an open pond where the plants can use the maximum amount of the sun's energy, a point can be reached where assimilation of minerals produces a scarcity which limits growth. At the other end of the scale, growth of plants at a depth of less than a metre can be brought almost to a halt due to the inability of the sun's rays to penetrate to that depth. This may be caused by a thick covering of floating weed or turbulence due to a large amount of suspended organic matter. A farm pond richly supplied with organic material draining into it from a near-by manure heap, can support a massive growth of the flagellata alga, *Euglena viridis*. Even though the water is shallow, little light can penetrate and growth of other weeds is impossible.

Oxygen

The presence of a large amount of weed growth can affect the oxygen balance in a pond only in providing oxygen by photosynthesis but it can have the opposite effect for the ultimate decay of the crop may create an oxygen demand theoretically equivalent to the oxygen gain during the growing period.

In a stream, large growths of weed can have important physical effects by reducing the velocity and increasing the depth. This may give rise to silting and flooding in other parts which, in turn, can mean the retention of decomposing organic matter in a short stretch of the stream, thus affecting the oxygen balance in that region.

Buoyancy and structural adaptations

Another factor affecting aquatic plants is the high density of water (700 times that of air at s.t.p.). This increases buoyancy so that rigidity, important to land plants subject to windy conditions, is not necessary and their stems do not require so much supporting tissue. The reduction in woody tissue becomes evident if a submerged plant such as Water starwort (*Callitriche platycarpa*) is lifted out of the water. In air its stems are totally unable to support the plant in an upright position.

Plants which grow wholly or partly under water show other characteristics. The shape of their submerged leaves is often narrow and strap-like thereby offering less resistance to water currents and wave action. *C. platycarpa* (figure 7) has a rosette of floating leaves at the surface which help to support the flimsy stem, while the submerged leaves are narrow and pointed.

An extreme example of leaf reduction is seen in the River crowfoot (*Ranunculus fluitans*) (figure 7). It is normally found growing in strong river currents with its thread-like leaves attached to long trailing stems which offer little resistance to the water flow. On the

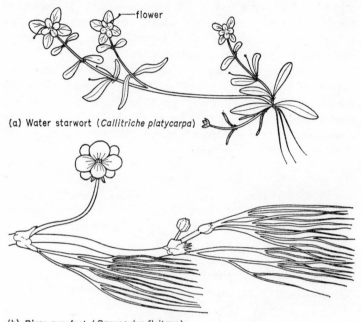

(a) Water starwort (*Callitriche platycarpa*)

(b) River crowfoot (*Ranunculus fluitans*)

(c) Water crowfoot (*Ranunculus aquitilis*)

Fig. 7.

other hand Water crowfoot (*R. aquatilis*) (figure 7), typical of
ponds and slow-flowing water, has leaves of two types. The underwater leaves are finely divided while the floating leaves are broad
and palmate as in its terrestrial relatives.

Differences in leaf shape on a single plant are well illustrated
in those of Arrowhead (*Sagittaria sagittifolia*) (figure 8), in which
the lowest leaves which are subject to underwater currents, are
narrow and grass-like. Ovate leaves develop next as the plant
grows towards the surface while the emergent leaves, last to
develop, are stiff and arrow-shaped.

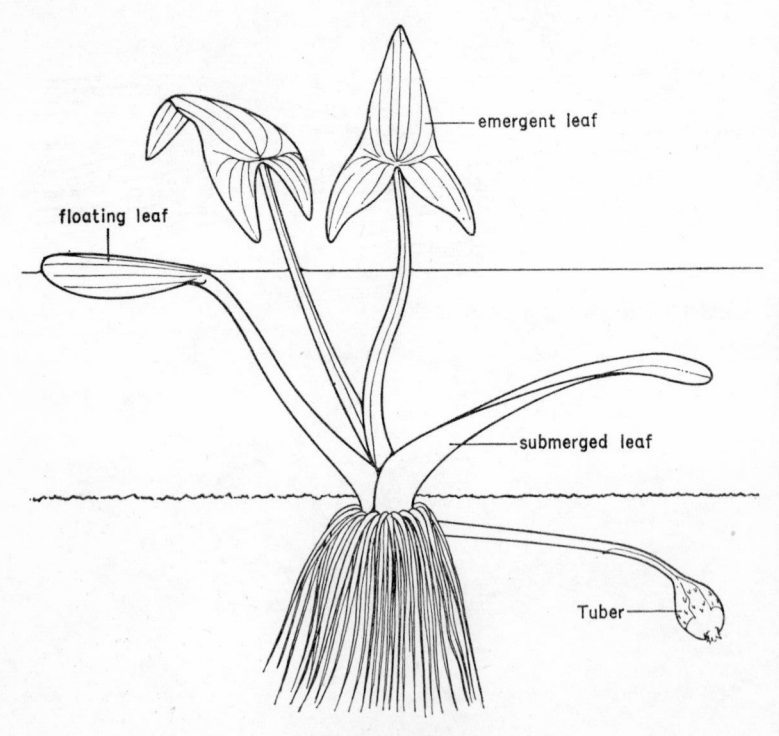

Leaves of Arrowhead (*Sagittaria sagittifolia*)

Fig. 8.

Plate 1 Western bladderwort (*Utricularia neglecta*)

(*a*) Rootless submerged frond showing much-divided leaves bearing bladders. (*b*) Portion of frond (× 35) with bladders.

Plate 2 Water soldier (*Stratiotes aloides*)

(a) Plants with new foliage floating at the surface in early summer. A white flower can be seen on the plant in the middle foreground.

(b) Plants in late summer (John Clegg).

Plants such as the Water lily (*Nymphaea alba*) show another peculiarity. The stomata, through which gaseous exchange takes place, unlike those of land plants, are mostly confined to the upper surfaces of the leaves so that they can make direct contact with the air. Their position can be demonstrated by submerging a water lily leaf in a bowl of water. By blowing hard through the cut end of the stem, small bubbles will be seen escaping through the stomatal apertures on the upper surface.

Pollen produced by terrestrial plants is transferred from one flower to another usually either by wind or by insects. Pollination can take place in the same way in aquatic plants which have aerial flowers, but for those which normally live beneath the surface special devices for successful pollination are required. Some, such as Amphibious persicaria (*Polygonum amphibium*) and Floating pondweed (*Potamogeton natans*), produce floating leaves which act as a platform for the support of the flowering spikes (figure 9 (a)). In the Water violet (*Hottonia palustris*) (figure 9 (b)) a thick mat of finely divided leaves grows just beneath the surface while the flowering stem, bearing whorls of pale mauve flowers, arises from the centre of the leaf mat.

In more extreme cases where there are no floating leaves and the plant remains totally submerged, special devices are employed to bring the flowers to the surface. *E. canadensis* rarely flowers. Occasionally, however, a long thread-like flower stem, often as much as 15 cm long, arises from the axil of a leaf. At the end is a small, very pale, pink female flower. Male flowers are rare but when they are produced, buds float up to the surface to liberate pollen which blows along the water to pollinate the female flowers. The sexual method of reproduction in *E. canadensis* is uncommon; more usually the plant reproduces vegetatively, portions of the stem breaking away to form new plants.

Vegative reproduction is common among most water weeds and many produce special structures called *turions*, by means of which they overwinter. Turions are really specialized stems shortened and compact, so that the leaves are packed down close on top of one another. In *E. canadensis* the leaves at the tips of the

B

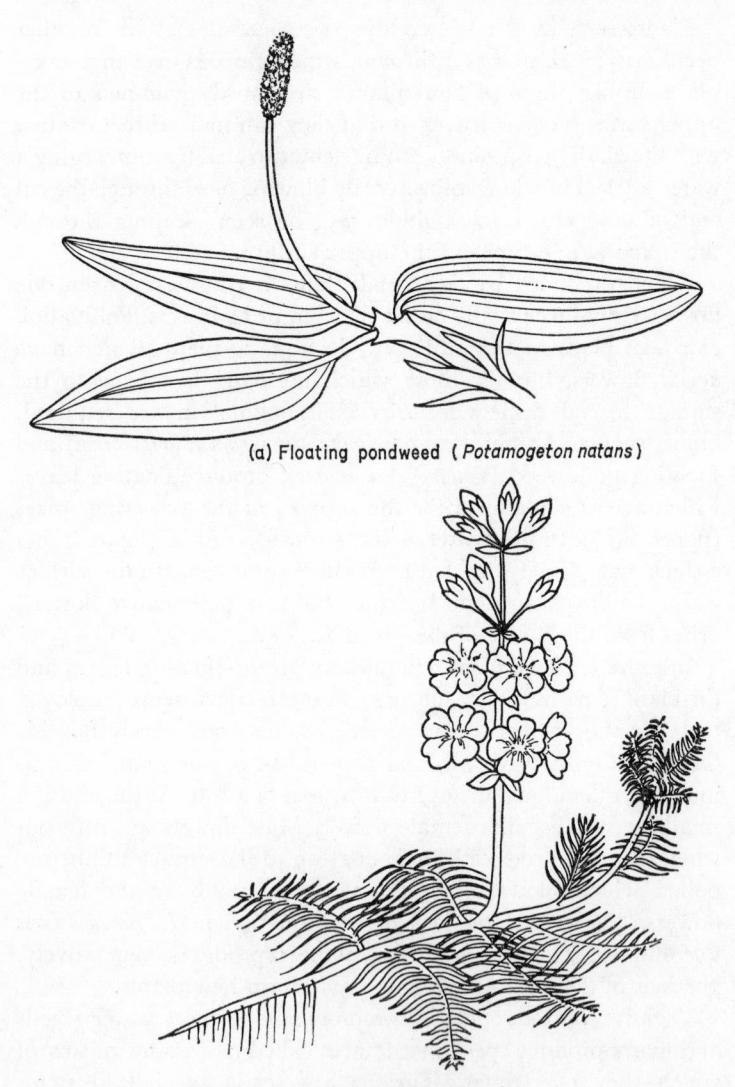

(a) Floating pondweed (*Potamogeton natans*)

(b) Water violet (*Hottonia palustris*)

Fig. 9.

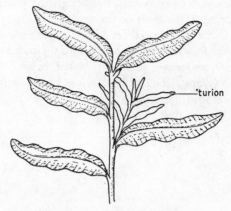

(a) Curled pondweed (*Potamogeton crispus*)

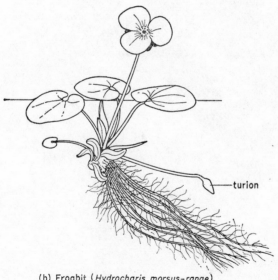

(b) Frogbit (*Hydrocharis morsus-ranae*)

Fig. 10.

stems become more compact although the parent plant persists all winter and the buds do not separate.

The Curled pondweed (*P. crispus*) produces turions in the form of side shoots which break away from the parent plant and grow fresh roots and leaves (figure 10 (a)). Frogbit (*Hydrocharis morsus-ranae*) develops tight bud-like turions on the ends of long under-water stolons (figure 10 (b)).

3

The animal inhabitants of a pond

A pond can be described as a body of still water which is sufficiently shallow to enable attached water plants to grow all over it.
This cannot hold true for all ponds since we have already seen
that a cattle pond only a third of a metre deep, can be totally
devoid of rooted plants because of numbers of small organisms
living in the water which diminish the amount of light. For our
purposes, however, a pond can be taken to mean a small area of
water with vegetation extending to the middle.

The plants of a pond grow in zones. Marsh plants, or *hygrophytes*, form a zone around the edge of the pond. Rooted plants
with emergent leaves and floating plants are found in the deeper
parts and are called true aquatics, or *hydrophytes*.

Animals, although they may have close associations with the
plant zones, are free to move about and must be grouped in a
different way.

Each of the physical areas of a pond described in this chapter,
imposes restrictions on the animals living there, for each species
exhibits certain characteristics which adapt it to special living
conditions. The most important of these adaptations are concerned with how the animals breathe, move, and obtain their food.
Their rate of respiration is dependent upon the amount of oxygen
present, each species having its own oxygen requirements. Modifications of structure of the respiratory organs range from practically no change from similar structures found in their terrestrial
counterparts, to extreme adaptations involving the evolution of
complex gills. There are also modifications of body shape or limb
structure which enable a species to make use of the water for
locomotion.

Obtaining food may mean the creation of a feeding current either by movement of cilia as in the freshwater cockles or by special limbs like those of the water fleas. Actively seizing the prey may mean the development of jaws and other mouthparts such as those of the carnivorous water beetles. Freshwater carnivores which have to hunt their prey, may do so passively and by stealth like nymphs of dragonflies, or actively by the fast movements typical of predatory animals such as the Water boatman (*Notonecta glauca*) and the Three-spined stickleback (*Gasterosteus aculeatus*).

The surface film

If a steel sewing needle is carefully lowered on to the surface of water in a beaker, it will float, supported by the surface film. A heavier object will break through the film and sink. A few small animals, mostly insects, are so light that they can exploit the surface tension and walk or skate about without penetrating it. Not only does their lightness enable them to do this but their bodies and especially their tarsi, are covered with *hydrofuge* or water-repelling hairs.

Several species of water bugs, belonging to the order *Hemiptera*, are surface dwellers. Figure 11(a) shows the Water measurer (*Hydrometra stagnorum*), an excellent example of an insect which lives completely on the surface, never penetrating the water beneath. By observing *H. stagnorum* from water level as it swims in an aquarium, it can be seen that the body remains raised above the surface supported by the tarsi of the second and third pairs of legs, while the first pair is held out in front of the head, ready to grasp and devour any small insects falling on the surface. Piercing and sucking mouthparts, typical of the *Hemiptera*, are used to secure the prey, and the elongated head may be significant in giving the extra mobility required. Push the insect beneath the surface and a silvery covering to the body can be seen which indicates that a layer of air is trapped in the hydrofuge hairs surrounding the body. On releasing it again the insect will immediately regain its position on the surface.

Pond skaters (*Gerris* spp.), and water crickets (*Velia* spp.) are other insects which use the same method of locomotion. All these surface dwellers take in air through spiracles opening into the tracheal system, in the same manner as their terrestrial relatives.

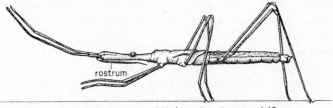

(a) Water measurer (*Hydrometra stagnorum*) 10mm

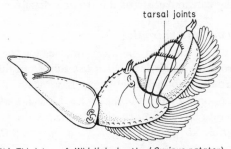

(b) Third leg of Whirligig beetle (*Gyrinus natator*)

Fig. 11. (a) The body of the Water measurer (*Hydrometra stagnorum*) is light and covered with hydrofuge hairs which prevent it from sinking through the surface film. (b) The second and third pair of legs of the Whirligig beetle (*Gyrinus natator*) are flattened and heavily fringed with hairs enabling the beetle to swim well both on the surface and beneath it.

The Whirligig beetle (*Gyrinus natator*) is well adapted for moving on the surface in swift girations. The body is streamlined and the second and third pair of legs are modified as paddles, each segment being flattened and heavily fringed with hairs (figure 11(b)). As the legs are pushed backwards in a swimming stroke, the tarsal fan is spread out, offering greater resistance to the water. On the return stroke the fan is folded. The front pair of legs are capable of grasping floating prey and the eyes are divided into two parts, the upper capable of vision above the surface and the lower beneath it.

While on the surface, *Gyrinus* obtains its supply of air direct from the atmosphere. But the beetle can also dive to avoid danger. In doing so, an air supply is carried underneath the wing cases and in a bubble attached to the end of the abdomen.

Open water

Many species live in the water between the surface and the bottom To this group belong animals which breathe air and which must therefore visit the surface at intervals to renew their supply. There are also those which possess gills of different kinds by means of which they can make use of the oxygen dissolved in the water.

Partially aquatic animals

Animals which have no special structures for the extraction of oxygen from the water but which use that from the air, are said to be partially aquatic. They must have ways of reaching the surface and, once there, of breaking the film so that their spiracles can come into direct contact with the air. At the same time, they must be able to prevent water from entering the respiratory system.

Most of the aquatic beetles, many insect larvae, and some molluscs are partially aquatic. In the Great diving beetle (*Dytiscus marginalis*) the last pair of legs is flattened and equipped with a long fringe of hairs which enables the beetle to swim powerfully (figure 12(a)). The beetle visits the surface at intervals to renew its air supply and to do this it turns hind-end uppermost breaking the surface film with the hydrofuge hairs at the tip of the abdomen. The wing cases are slightly raised and the air concealed beneath them is replenished. Each abdominal segment bears a pair of spiracles on the dorsal surface. In taking in air, the two spiracles on the last segment are raised above the surface and water is prevented from entering by the circlet of hydrofuge hairs at the tip. The rest of the spiracles renew their supply from the air lodged beneath the wing cases and this can take place while the beetle is submerged.

The air carried under the elytra in *Dytiscus* is used exclusively as a reserve upon which the beetle draws when it is submerged. If the membraneous wings are removed, the space beneath the elytra is increased. Such specimens can remain submerged for a longer period.

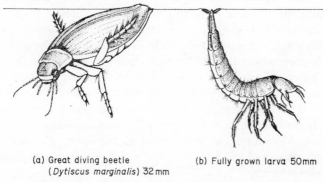

(a) Great diving beetle   (b) Fully grown larva 50mm
(*Dytiscus marginalis*) 32 mm

Fig. 12. Both the adult (a) and larva (b) of the Great diving beetle (*Dytiscus marginalis*) are partially aquatic, obtaining air at the surface.

In the case of *Dytiscus* the amount of air which it normally carries is sufficient to allow it to float up to the surface by its own buoyancy. It must then swim actively down from the surface and, in order to remain below, must continue to swim or to cling to a submerged object.

The larva of *Dytiscus* (figure 12(b)) is also partially aquatic. To renew the supply of air it swims as if walking in the water, using its long fringed legs. The two cerci at the end of the abdomen break the film and air is drawn into spiracles lying between the cerci.

In other partially aquatic insects air is carried on the outside of the body sometimes beneath a canopy of hydrofuge hairs on the ventral surface of the abdomen as in the Water boatman (*N. glauca*) or as a bubble of air at the end of the abdomen as in some of the smaller dytiscid and hydrophilid beetles. In these cases the superficial air bubble, or '*plastron*', is in direct contact with the water and functions as a 'physical gill'. When submerged, these insects withdraw oxygen from the air bubble for their physical requirements. This reduces the oxygen within the bubble until a point is reached when the partial pressure of this gas is less than that of the surrounding water, oxygen will then enter the bubble by diffusion. Such a reduction in the oxygen tension would then be accompanied by a rise in the amount of carbon dioxide within the bubble were it not for the fact that this gas is readily soluble in water. This results in an increase in the amount of nitrogen within the bubble and when the partial pressure of nitrogen rises above that of the surrounding water, nitrogen will diffuse slowly out of the bubble. The nitrogen merely acts as a medium enabling oxygen to diffuse into the bubble but the rate at which this happens will decrease as the amount of nitrogen diminishes. When a critical level is reached, the insect will suffer an oxygen want and will be forced to rise to the surface to replenish its reserve of air.

Certain molluscs are also partially aquatic, visiting the surface at intervals to breathe air taken into a chamber beneath the mantle through a breathing aperture. The lining of the mantle is richly supplied with blood vessels and functions as a kind of lung, hence they are known as *pulmonate*. Any of the soft body tissues, such as the head or foot of the mollusc, which comes into direct contact with the water, are also able to absorb oxygen from it. The Great pond snail (*Limnaea stagnalis*) often uses the surface film for crawling, hanging upside down with the foot expanded and clinging to the film. To do this, extra buoyancy is achieved by filling the mantle cavity with air. One can often watch individuals that have imbibed a quantity of air, turning and twisting about in an endeavour to abandon the surface.

The lung aperture has completely disappeared in the fresh-water limpets, *Ancylastrum fluviatile* and *Ancylus lacustris*, and although they are pulmonate snails they never visit the surface and breathe entirely by diffusion through the skin. They are able to obtain sufficient oxygen because their surface area is great compared to the body volume, which allows an efficient gaseous exchange.

Many pulmonates, however, live in stagnant water containing little or no dissolved oxygen. Under such conditions respiration is restricted to the absorption of atmospheric air by the mantle cavity and the less oxygen that the water contains, the more frequently must they visit the surface. This is certainly true of *L. stagnalis*, yet when the pond is covered with ice and they are cut off from atmospheric air, they seek the muddy regions at the bottom and survive by cutaneous respiration in which the expanded cephalic tentacles play an important role (figure 13 (a)).

The planorbid snails such as the Great ramshorn snail (*Planorbis-corneus*) (figure 13 (a)), which have thread-like tentacles, possess the respiratory pigment haemoglobin in their blood which increases the amount of oxygen that the blood can absorb. Under conditions of oxygen shortage, the edge of the mantle projects and acts as a pseudogill. The lower the concentration of oxygen the more this part of the mantle projects.

Planorbids can exist in concentrations of oxygen which are much lower than that required by most freshwater molluscs. Janus (1965) mentions that at 15°C, *L. stagnalis* must renew its supply of air when the oxygen in the respiratory chamber falls to 13 per cent. At the same temperature, *P. corneus* does not need to visit the surface for air until the oxygen content of the respiratory chamber reaches 4 per cent. From this it is apparent that although under certain conditions of oxygen shortage some freshwater pulmonates can absorb dissolved oxygen through other parts of the body than the mantle cavity, the method of respiration they adopt is determined by their seasonal requirements as well as by the concentration of oxygen in the water.

Totally aquatic animals

To this group belong all the species which are dependent upon the dissolved oxygen in the water for their respiratory needs.

In small organisms such as the *Protozoa*, the possession of a large

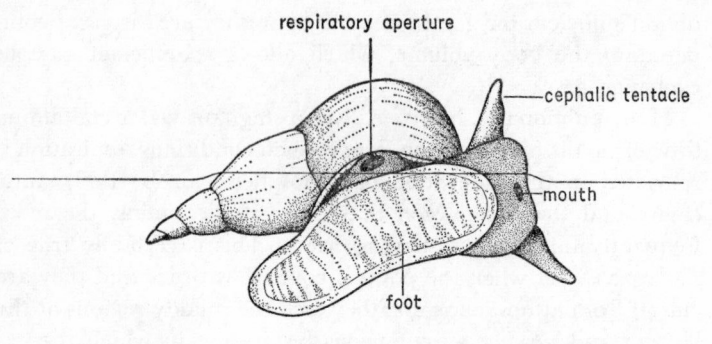

(a) Great pond snail (*Limnaea stagnalis*) 35mm high

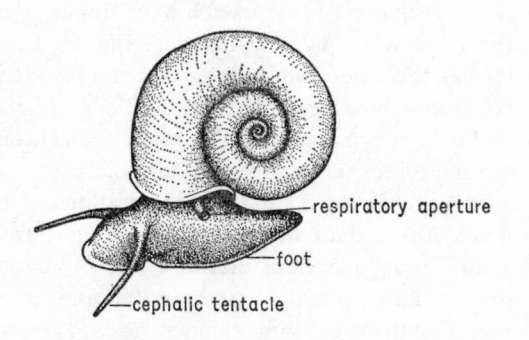

(b) Great ramshorn snail (*Planorbis corneus*) 25mm across

Fig. 13. Two pulmonate molluscs commonly occurring in stagnant water.

surface-area/volume ratio means that an effective exchange of gases can be maintained without the necessity of special respiratory organs. This system of gaseous exchange is usually accompanied by the possession of a thin and highly permeable skin. Even larger organisms such as leeches and other freshwater annelids are able to maintain an efficient exchange of gases through the body wall. The smaller crustaceans, too, such as water fleas and ostracods fall into this category for although their bodies have a hard exoskeleton, their limbs are modified in various ways to create a current of water which flows over the soft body parts. This also acts as a feeding current by wafting particles of food towards the animal. The water flea (*Simocephalus* sp.) (figure 14(a)) feeds and breathes in this way. The paired limbs are leaf-shaped and heavily fringed with hairs. Constant movement of the limbs draws a current of water through the carapace, and at the same time small particles of food, such as diatoms and other microscopic algae, are filtered off by the hairs on the limbs and are passed forward to the mouth. *Simocephalus* progresses jerkily through the water by the flicking movements of the long second pair of antennae.

Many of the two-winged flies have aquatic larvae, and some like the other small animals mentioned above, have no special respiratory organs. The larva of the gnat, *Chaoborus* sp. (figure 14(b)) besides being of interest because of its strange habits and structure, plays an important part in the food relationships within a pond. It feeds on small crustaceans, rotifers, and algae, seizing the prey with its antennae which are modified as prehensile organs. An elaborate tailfin, composed of a comb of bristles, acts as a rudder and is an aid in propelling the larva through the water as it swims by jerky side-to-side movements of the body.

As with all insects, the body of *Chaoborus* is covered with chitin, but this is so thin that a direct exchange of gases can take place. Air, or a gas of some kind, fills the paired bean-shaped sacs at either end of the body. These are hydrostatic organs which can expand or contract, in this way adjusting the buoyancy of the larva to the density and pressure of the water, permitting it to rise or fall.

Many totally aquatic species such as fish, have elaborate gills, the surface area of which is greatly increased by lamellar folds. At the other end of the scale there are the very simple gills possessed by the larvae of the midge, *Chironomus* sp. These are just outgrowths of the body probably merely serving to increase the body surface (figure 14(c)). Between these two extremes, there are a great variety of gill structures, often modified according to the place in which the animal lives. Basically a gill is an outgrowth

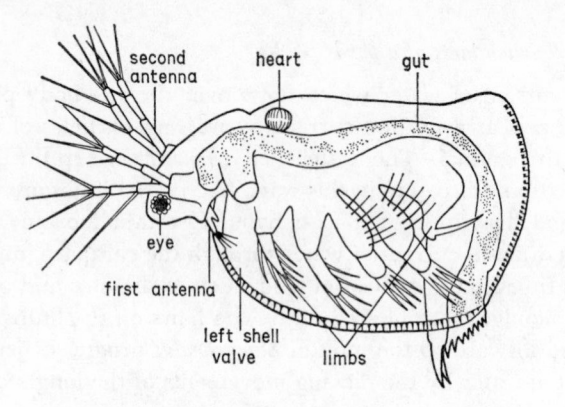

(a) Water flea (*Simocephalus* sp.) 3 mm

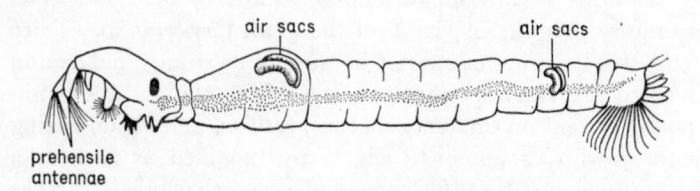

(b) Larva of the Phantom midge (*Chaoborus* sp.) 10 mm

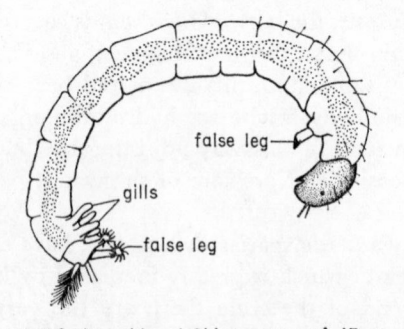

(c) Larva of the midge (*Chironomus* sp.) 15 mm

Fig. 14. Three examples of totally aquatic organisms. Movement of the paired limbs in *Simocephalus* (a) maintains a current of water, gaseous exchange taking place through the surface of the limbs. In both *Chaoborus* (b) and *Chironomus* (c) gaseous exchange takes place through the thin integument, although *Chironomus* also possesses simple gills.

from the body, through the surface of which a diffusion of oxygen and carbon dioxide can take place. In some filter-feeding organisms, such as the Freshwater mussel (*Anodonta cygnea*), the gills are covered with cilia which pass a string of mucus containing food particles, up the gill filaments. The gills in these cases perform both a feeding and respiratory role (figure 15).

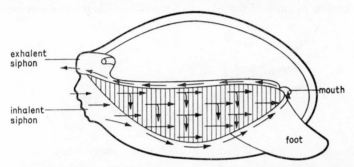

Fig. 15. Schematic drawing of the Swan mussel (*Anodonta cygnea*) with the right lobe of the mantle removed to show the outer lamella of the right gill. Water taken in at the inhalent siphon passes over the gills. The vertical arrows represent ciliary currents passing down the gill, those at the bottom of the gill indicate the direction of the main food current running towards the mouth, while those at the top of the gill indicate the direction of the current passing to the exhalent siphon.

Very few adult insects are totally aquatic but there are many species whose larvae and pupae have developed gills of different kinds.

The larva of the Mayfly (*Chloeon dipterum*), typical of still water, is to be found at most times of the year (figure 16). Down each side of the abdomen is a series of plate-like gills. Each pair except the last is double. The first six pairs are kept in constant vibration setting up a current of water which passes backwards until it meets the seventh pair of gills. These are kept stationary and deflect the current of water to each side so that it will not immediately be used again. In water highly charged with oxygen, movement of the gills is slow. But if the nymphs are transferred to pond water which has been boiled and cooled to the original temperature, the gills immediately start to vibrate much more rapidly in an endeavour to waft oxygen towards them.

Dragonflies belonging to the order *Odonata*, all have nymphs which are totally aquatic. Two different methods of breathing are used by the nymphs and this serves to separate the species into two distinct groups. To the first belong the large and sluggish fat-bodied nymphs of the hawker dragonflies. They live on or near

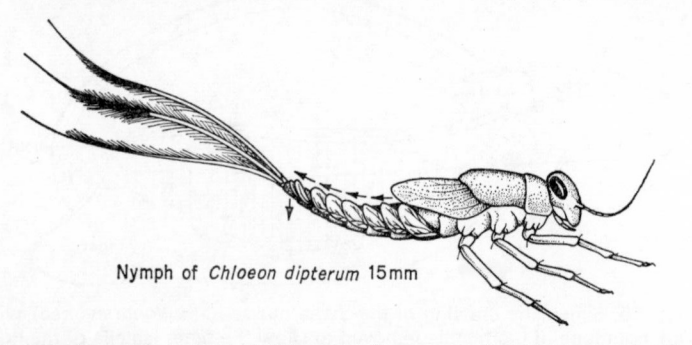

Nymph of *Chloeon dipterum* 15mm

Fig. 16. A nymph typical of still water. A current of water created by the constantly moving gills, flows over the body but is deflected by the stationary seventh pair of gills so that it will not immediately be used again.

the bottom and being a dull shade of green, resemble their general background colour. They can remain motionless for long periods, relying on stealth when stalking their prey which is seized by a structure called the mask (figure 17 (a)). This is really the fused third pair of jaws ending in strong hooks. At rest the mask is kept folded underneath the head, but it can be shot forwards to grasp the prey. The whole mask is then pulled back to bring the food to the mouth.

In the Hawker dragonfly nymphs the gills occupy a strange position, for they line the rectum, water being pumped in and out of the anus by contractions of the abdomen. This is also the means of rapid movement for when the water is expelled from the anus, the nymph is jet-propelled forwards.

The second group of dragonfly nymphs are those of the damsel-flies (figure 17 (b)), which have long slender bodies. They also seize

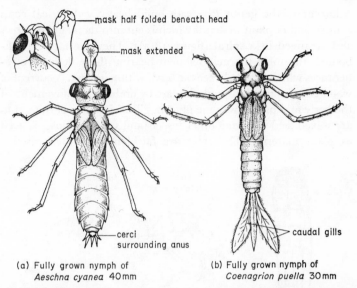

(a) Fully grown nymph of
Aeschna cyanea 40mm

(b) Fully grown nymph of
Coenagrion puella 30mm

Fig. 17. (a) The heavily built, sluggish nymph of a Hawker dragonfly with internal rectal gills and (b) a delicate and more active Damselfly nymph with three leaf-like gills.

their prey with a mask, but they are much more active, swimming about among the pondweeds by side-to-side movements of the body assisted by three flattened tail lamellae. The lamellae contain a number of tracheae and are usually considered to function as gills, although if they are accidentally lost, the nymph still lives, apparently able to respire perfectly well through the body wall. In fact the damselfly nymphs occupy much the same position in a pond as the nymphs of *Chloeon* and although the gills are different, their body shape and movements are similar.

The caddis flies (order *Trichoptera*), all have aquatic larvae and pupae. Their mouthparts, and particularly the mandibles, are modified according to whether they are carnivorous, herbivorous, or mixed feeders. The arrangement and shape of their gills also varies. The larvae of two genera which typically inhabit ponds, are described.

Larvae of the genus *Phryganea* build heavy, cylindrical cases
constructed of plant material which is cut into rectangular pieces
and arranged in a spiral (figure 18(a)). They are fat and soft-
bodied, the third thoracic segment bearing three fleshy protru-
berances which help to brace the larva within its case. A current of
water is maintained through the case by undulatory movements of
the body and passes over the filiform gills. Members of this genus
are mixed feeders eating other insects and small molluscs as well
as plant material. The cases are fairly heavy which restricts

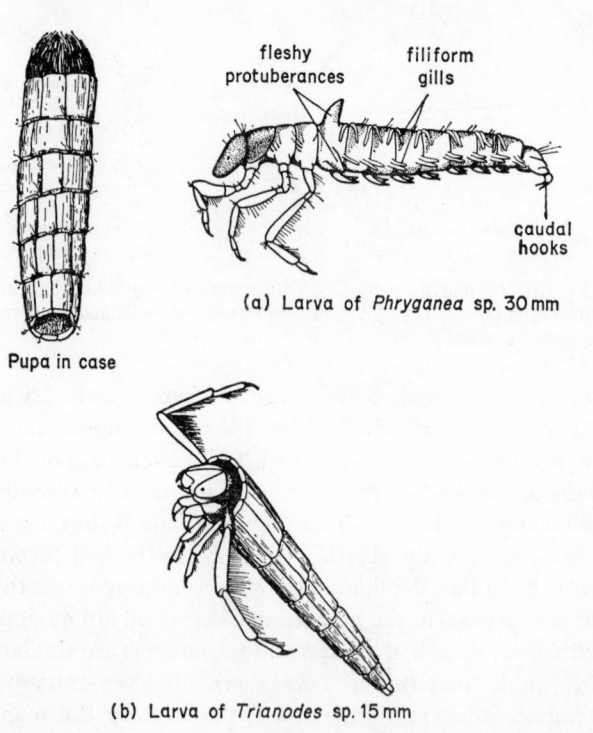

Pupa in case

fleshy   filiform
protuberances   gills

caudal
hooks

(a) Larva of *Phryganea* sp. 30 mm

(b) Larva of *Trianodes* sp. 15 mm

Fig. 18. Two caddis larvae commonly found in still water. (a) *Phryganea*
and its heavy cylindrical case inside which it pupates, blocking the entrance
with plant fragments. (b) *Trianodes* inside its much lighter case. Head and
thorax project from the case as the larva swims by means of its third pair
of legs, heavily fringed with hairs.

movement to crawling and the larva pupates within its case by blocking up the ends with plant fragments.

Larvae of the genus *Trianodes* build much lighter cases constructed of plant material arranged in a very even spiral to form a long conical structure from which the head and thorax of the larva protrudes (figure 18 (b)). The body of the larva is similar to that of *Phryganea* but the third pair of legs which are long and fringed, are used for swimming, propelling the larva, case and all, through the water.

In contrast to the pulmonate molluscs, none of which possess gills, there are some which have a horny plate, called the *operculum*, attached to the foot. This fits the shell opening closing it completely when the animal is withdrawn. These snails are called operculates and are found in reasonably well-aerated water, breathing dissolved oxygen by means of special gills. In one species, the Common valve snail (*Valvata piscinalis*), the gill is particularly evident (figure 19(a)).

Several species of bivalve molluscs, the orb-shell cockles (*Sphaerium* spp.) and the pea-shell cockles (*Pisidium* spp.), are very common in some ponds (figure 19(b)). The mantle is produced into two long siphons in the former or one shorter one in

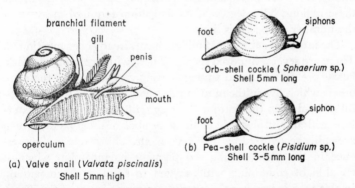

(a) Valve snail (*Valvata piscinalis*)
Shell 5mm high

Orb-shell cockle (*Sphaerium* sp.)
Shell 5mm long

(b) Pea-shell cockle (*Pisidium* sp.)
Shell 3-5mm long

Fig. 19. (a) *Valvata piscinalis*. A gill-breathing gastropod with an operculum. The foot is deeply cleft at the front and the penis is non-retractile. (b) Two genera of small bivalve molluscs commonly found at the bottom of ponds.

the latter, through which water is drawn into the gill chamber. The lamellar gills are ciliated and the water circulating within the gill chamber is used not only for respiration but also to bring in particles of food which are trapped in the mucus of the gills. The cilia propel the food mucus string towards the mouth. These small cockles move about in the mud by means of a very extensible worm-like foot. Calcium is necessary for all species of freshwater molluscs for shell building. Some require more than others. Plants which thrive in calcium-rich water provide their food and for gill-breathing molluscs the presence of lime in the water is especially important since it causes precipitation of any clay particles, thus preventing the water from becoming turbid.

The inhabitants of the mud

The regions of a pond so far described include the surface and the open water, both of which are places in which there can be plenty of dissolved oxygen. But in the mud at the bottom conditions can be very different for often there is a great accumulation there of decaying plant and animal matter. This means that oxygen is used up in the process of decay and where there are heavy deposits there may be a complete lack of oxygen. Even under such conditions there are a few animals which have been able to exploit these regions. The degree of specialization required means that the number of species is limited.

The larvae of chironomid midges have already been mentioned. There are several hundred species. Those with green or colourless larvae are usually found among floating vegetation, but there are others which live in the mud in tubes constructed of silt. A scoopful of mud from the bottom of a stagnant pool or from a water-butt will almost always contain a number of these larvae which are easily recognized by their red colour. This is due to the presence in the blood of the red pigment haemoglobin which is capable of combining with oxygen to form oxyhaemoglobin, in this way increasing the oxygen-carrying powers of the blood. This enables the larvae to live in mud, often rich in decaying organic matter, where oxygen is not so plentiful.

The annelid worms, belonging to the genus *Tubifex*, are also inhabitants of the mud. Their blood, like that of *Chironomus*, contains haemoglobin. They live, head downwards, in tubes constructed of mud and wave the posterior ends of their bodies ceaselessly. These continuous waving movements create a current which enables the worms to extract oxygen from a wide area of water. They do not possess special gills, oxygen being absorbed over the general body surface. If the concentration of this gas decreases, the worms protrude further out of their tubes, thereby exposing a larger absorbing area of the body.

4

Factors affecting the distribution of freshwater organisms

The distribution of organisms in fresh water depends upon their physical requirements relative to their environment. In addition, a high rate of reproduction combined with effective powers of dispersal are an advantage both in reaching a new habitat and in becoming established there. Once having arrived, their continued survival will depend upon the availability of food and their ability to shelter from predators.

Chapter 2 examined the environment of fresh water from the point of view of its chemical and physical properties which in terms of the ecologist are really factors influencing the distribution of the organisms living there. How these operate can best be illustrated by taking a few examples from different freshwater habitats. It is as well to realize that distribution maps can be most misleading, for they can easily reflect the distribution of the recorders rather than that of the organisms being recorded! This can be especially true in the case of reputedly rare species, identified by the field naturalist who is not necessarily an expert taxonomist.

Investigation of any area of fresh water will soon show that some species are common while others are rare. The next question is whether the rare species have just arrived, so to speak, and are in the process of establishing themselves within the community, or whether the environmental conditions are such that their numbers have decreased to the point of scarcity.

Some species have greater tolerance of a wide range of conditions than others. The copepod, *Cyclops fimbriatus*, for example, is to be found in ponds of varying temperature and mineral content all over Europe from sea level to 3000 m. On the other hand, another crustacean, *Triops* (*Apus*) *cancriformis*, is only found in a few isolated colonies. In Great Britain it has been recorded from widely separated counties such as Hampshire, Kent, and Kirk-cudbrightshire, usually in temporary pools where a necessity for its existence seems to be a periodic drying out of the mud. The ability to survive long periods of desiccation is associated with the fact that *Triops* produces only resting eggs, which may or may not be fertilized, males of the species being very rare.

Among the molluscs, the Wandering pond snail (*Limnaea pereger*) and the Great pond snail (*L. stagnalis*) are fairly ubiquitous while another species of the same genus, *L. auricularia*, has a more restricted range, preferring large bodies of water with a higher concentration of calcium.

Brackish water as a transition area

Organisms which are endeavouring to find a footing in a new environment, can best be studied in boundary areas between one kind of habitat and another. Brackish water, existing between estuary and fresh water, is one such area, although it should be realized that the distinction we make between sea water, brackish water, and fresh water is merely one of convenience. To an adapt-able species, the changes from sea water to fresh water are con-tinuous and the term *mixohaline* is often used to describe brackish water or dilute sea water.

The salinity of sea water is roughly 35 g of salts per litre of water, but this, of course, varies in different parts of the oceans. In the Red Sea, for example, due to heavy evaporation the salinity may be as high as 40 g/l while in the Baltic, renowned for its variation in salinity, it can be as low as 2 g/l.

Where fresh and sea water meet, species from each invade the mixohaline region. A few species, typical of fresh water, can combat the violent changes of current as well as the salinity consequent upon tidal action, spreading, though in diminished numbers, into the mixohaline regions. Several species of *Hydro-bius*, *Platambus maculatus*, and *Deronectes depressus* are examples of smaller beetles, while a few other insects such as some species of *Notonecta* and *Corixa* are also often found. Soft-bodied animals such as planarians and many of the segmented worms and all the *Cladocera* are rarely found in mixohaline waters, the rapid diffusion of salts through the thin body covering of these organisms is probably the reason.

In some areas of salt marshes where fluctuations in salinity are great, there may be periodic invasion by typically freshwater organisms. Nicol (1935) investigating pools in a salt marsh found that the salinity of those reached only by spring tides, was about 1 g/l but that after they had been invaded by the tide, the salinity rose to 17·5 g/l. Six species of water beetle, found usually in fresh water, were inhabitants of these pools, but the rest were typical mixohaline species and included *Gammarus duebeni* and the mosquito, *Aedes detritus*. During the neap tide periods when the salinity fell once more, several typically freshwater species, such as the mosquitoes *Culex pipiens* and *Anopheles claviger* as well as the snail, *L. truncatula*, invaded the pools but these did not survive innundation.

The activity and rate of development of a species, and thus its distribution, may be affected by changes in salinity. Schlieper (1958) showed that this was the case in the oyster, *Ostraea virginica*. At low salinities the activity and the rate of development of the mollusc was slowed down below the point at which spread was possible. Results at a temperature of 18–21·5°C were as follows:

35–25 g/l	development normal
23–21 g/l	development normal but slow
19·3 g/l	development slow with some mortality
17·5 g/l	some eggs died, few larvae survived to the stage of shell production
15·8 g/l	development very slow and very few larvae reached the stage of shell production.

Although mixohaline waters have attracted the attention of physiologists because they are regions in which differences in the distribution of the organisms can be related to the one obvious factor, salinity, this may not be the sole factor. *G. duebeni* can tolerate a medium between fresh water and salt water of as much as 50 g/l salinity. At this salinity and at a temperature above 20°C development is greatly retarded and consequently its ability to spread. A rise to 22°C is fatal. Therefore, temperature as well as salinity affects the range of this species especially in shallow estuaries or ditches where there can be considerable fluctuations of temperature.

Those species which have been able to colonize brackish water, have all managed to achieve to a greater or lesser extent, osmotic

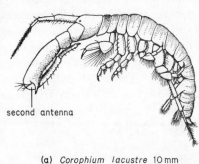

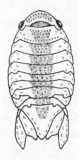

(a) *Corophium lacustre* 10 mm   (b) *Sphaeroma* sp. 10 mm

Fig. 20. (a) *Corophium lacustre.* An amphipod often present in large numbers in brackish water. (b) *Sphaeroma.* An isopod, also typical of semi-saline water.

independence and are termed *euryhaline.* A few organisms avoid to some degree fluctuations in salinity by burrowing into the mud.

Semi-tidal ditches are to be found at the head of an estuary such as the River Exe (Plate 3). The amphipod, *Corophium lacustre* (figure 20 (a)) is present in large numbers burrowing into the muddy sides of the ditches where it feeds on detritus. Equally abundant is another amphipod, *G. duebeni*, already mentioned. This species extends, although in diminished numbers, to the upper fresh-water reaches of the ditches. Other crustaceans, enjoying a degree of osmotic control, such as the burrowing isopod, *Sphaeroma*, and the prawn, *Palaemonetes varians*, are also present (figure 20 (b)).

Among the plants found in these ditches, the marine alga, *Enteromorpha intestinalis*, is found up to the limits of the brackish water, enduring periods of lower salinity when the tide is out. Water starwort (*Callitriche platycarpa*) flourishes in both the upper and lower ditches, along with other typically freshwater plants such as the Common rush (*Juncus communis*), the Common reed (*Phragmites communis*), and the Yellow flag (*Iris pseudacorus*).

The distribution of the species of *Gammarus* in tidal and estuarine waters, is interesting and according to Green (1961) they are encountered in the following order from marine to freshwater areas: *G. locusta*, *G. zaddachi salinus*, *G. zaddachi zaddachi*, *G. duebeni*, and *G. pulex*. *G. pulex* possessing blood with a low osmotic pressure, is confined to fresh water but a study of distribution of this species and of *G. duebeni* has shown that the latter has only become established in fresh water in which *G. pulex* is absent, being unable to compete with *G. pulex*. This is certainly the case in western regions of Britain. Elsewhere *G. duebeni* is restricted to mixohaline waters in which *G. pulex* is established at the fresh-water end.

Food and predation

So far as animals are concerned, a sufficiency of their natural food is a good reason for their presence within a community. In general, the more highly selective an animal is with regard to its diet, the more restricted will be its range of distribution. Conversely, an organism such as the larva of *Chaoborus* which is a carnivore feeding on a variety of zooplankton, is to be found in most bodies of fresh water.

Pacaud (1949) performed an interesting experiment using a culture of the Water flea (*Daphnia obtusa*) and the Great ramshorn snail (*Planorbis corneus*) which he kept in water to which nothing but cigarette paper was added. The snails fed on the cigarette paper and the water fleas upon the bacteria that decomposed the faeces of the snail. This culture was kept going for eight months. In a similar culture where the snails were absent, the water fleas all died within the space of seventeen days because they were unable to consume the bacteria decomposing the vegetable matter of the cigarette paper. Not only does this experiment show that the feeding habits of one species can affect the population of another but also that the food requirements of one of them (*D. obtusa*) are very specific.

The need for food often results in curious patterns of behaviour which control the size of the population of an animal. For instance the flatworm, *Crenobia alpina*, found in slow-flowing streams, feeds on carrion of various kinds. When food is scarce in one part of a stream it will go downstream, often in large numbers, to regions where food is more plentiful. Any excess food ingested is devoted to egg-production which results in a physiological change causing the animal to become positively rheotropic and to return upstream. If there is insufficient food in the upper part of the stream it starts to reabsorb its eggs and to become once more negatively rheotropic.

Calcium

Of all the mineral salts present in fresh water, calcium is probably the most variable in amount and the one which has the greatest effect on the distribution of freshwater organisms. Little seems to be known, however, of exactly how this comes about although the presence of calcium does facilitate the decomposition of organic matter and in this way it may well have an indirect effect.

It is possible to make a very rough division into two groups of those organisms which prefer soft waters and those which prefer hard waters. To the latter group can be relegated the molluscs and crustaceans on the ground that their shells are largely composed of lime and to the former, planarians and annelids. This is, however, a very broad division which, when the distribution of individual species is considered, does not really hold good. Among the molluscs, *Limnaea glabra* is characteristic of water with a low calcium content. *Planorbis carinatus*, typically a hard-water species, according to Macan (1963) is common in Lake Windermere with a low calcium content of 5 mg/l. In this case and probably in other molluscs also, a large volume of water seems to compensate for a low calcium content.

Reynoldson (1961) working on the occurrence of *Asellus* in a number of freshwater lakes came to the conclusion that the concentration of calcium and of dissolved organic matter affected its distribution. He rarely found the species in water with a concentration of calcium below 5 mg/l, that its appearance in water with a calcium content of between 7 and 12·5 mg/l was sporadic while it favoured a calcium content greater than 12·5 mg/l.

The same author (Reynoldson, 1958) collected planarians from 122 lakes with a wide range of calcium concentrations. The results he obtained (figure 21) seem to show that some species of triclad are distributed according to whether the water is hard or soft. While *Polycelis nigra* is tolerant of a wide range of calcium concentrations.

Temperature

A few freshwater invertebrates such as the flatworm, *Crenobia alpina*, typical of upland streams, and the amphidod, *Mysis relicta*, (figure 22) are relics of the Ice Age and are restricted to living within a relatively narrow range of temperature. *M. relicta* lives in the lower regions of certain lakes but should there be any enrichment of the lake bed in summer, causing a fall in the oxygen concentration, the animal can be exterminated altogether. For this reason, it is now confined in Britain to a few deep lakes. *C. alpina* avoids higher temperatures by moving upstream during summer to the cooler sources in the higher regions.

Although temperature is an important factor in the lives of many freshwater animals, its effect may be indirect rather than direct since temperature changes alter the amount of oxygen in solution which may well be the deciding factor in the cases just mentioned.

To breed successfully and to produce sufficient numbers of offspring, is one of the requisite factors in the establishment of a population. But many organisms have a

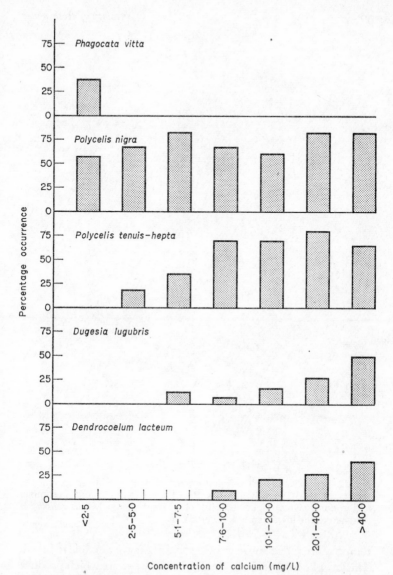

Fig. 21. Percentage occurrence of five species of triclad in waters with various concentrations of calcium. (After Reynoldson, 1958.)

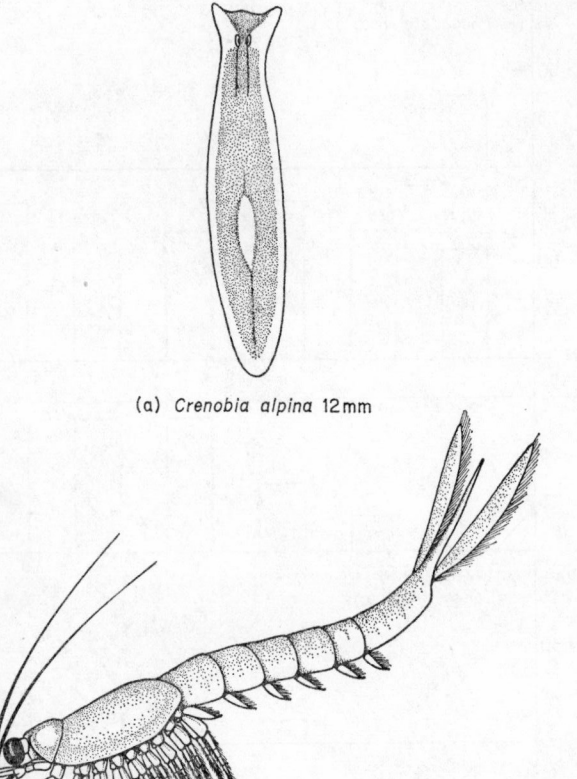

(a) *Crenobia alpina* 12mm

(b) *Mysis relicta* 5mm

Fig. 22. Two relics of the Ice Age restricted to living within narrow limits of temperature. (a) The planarian, *Crenobia alpina* moves upstream in summer to the cooler water higher up. (b) The Mysid shrimp (*Mysis relicta*) is confined to the lower regions of certain lakes.

temperature range above or below which they are unable to breed. This may be important especially for those species which produce more than one generation a year. The blackfly (*Simulium* sp.) lays its eggs on stones in swift-flowing streams. A first batch of eggs may give rise to numerous larvae early in April when the water is relatively cool. The chief predators of the larvae are plecopteran and ephemerid nymphs which usually make their appearance a month or so later, ready to attack the second batch of *Simulium* larvae.

A further difficulty is added to the whole question of temperature effect by the fact that although some species have a wide temperature tolerance, they probably do not occupy the whole range. This may be due to the presence of other species existing in competition, at the upper and lower limits of their range, and which have a different temperature optimum. At this temperature, the second species may breed faster or be more active and much better able to find and utilize its food.

Most of the work done on the effect of temperature on the distribution of organisms, concerns their rate of development and therefore their ability to increase and to colonize certain areas in which one species may be in competition with others. For organisms such as insects with a complicated life history, it is necessary to complete a stage in such a life history within a certain period of the year. In addition, it may be necessary to produce large numbers of offspring in order to compete successfully with a rival species or in order to maintain a population of sufficient size, in the face of predation.

A good example of the way in which temperature affects distribution is found among the *Dytiscidae*. Below 10 °C the young larvae of the Great diving beetle (*Dytiscus marginalis*) are too inactive to feed and therefore growth ceases. But between 11 °C and 15 °C growth increases rapidly. At a temperature above 27°C, the number of larvae failing to complete development increases steeply.

The larvae of a near relative, *D. semisulcatus*, living, feeding, and reproducing in the same regions and therefore in direct competition with those of *D. marginalis*, start to grow at 3 °C, a lower temperature than that necessary for growth of larval *D. marginalis*, and the amount of food they eat increases with rising temperature to 26°C. The larvae of this species are found throughout the winter while those of *D. marginalis* occur in summer and could not survive winter temperatures.

Combined effect of several factors

By now it must be evident that the effect of one factor is inextricable from that of another. Flowing water probably offers a greater complex of factors than does still water, for the speed of the current, temperature, and the amount of oxygen present are three factors which operate together.

It is difficult in the laboratory to equate conditions to those prevailing in a headland stream and there are bound to be discrepancies between experimental results and field observations.

Ambühl (1959) kept nymphs of various species of ephemerids on a substratum of powdered glass, in tubes through which flowed water containing 8 mg/l oxygen at 18–19°C. He found that in the case of *Ecdyonurus venosus*, an increase in current speed from 0·5 cm/sec to just over 6 cm/sec raised their oxygen consumption by less than 1 mg oxygen/g dry weight of body per hour, but for *Rhithrogena semicolorata*, the increase in oxygen consumption for the same increase of current speed was over 8 mg oxygen/g dry weight of body per hour. These differences in amount of oxygen consumption with rate of flow are due to the fact that in *R. semicolorata* the gills serve as a sucker for maintaining the position of the organism against being swept away by the current. The gills cannot, therefore, be used to create a current of water over the body. *E. venosus*, on the other hand, keeps the gills in constant motion thus maintaining a respiratory flow of water over the gill surfaces.

The net-spinning larva of the caddis, *Hydropsyche angustipennis*, a species found in fast-flowing streams, undulates its abdomen within its silken case at a rate which increases as the oxygen concentration falls. At a concentration of 8 mg/l oxygen, there is an increase of less than 1 mg oxygen/g dry weight of body per hour at a speed of less than 1 cm/sec and practically no increases at higher current speeds.

The effect of the combined influence of current, oxygen, and

Plate 3 Brackish dykes at Topsham, Devon, in July

(*a*) A dyke near its outflow into the estuary with Floating sweet grass (*Glyceria fluitans*) in the foreground. Yellow flag (*Iris pseudacorus*) and Soft rush (*Juncus effusus*) in the middle background.

(*b*) A dyke in the upper water meadows bordered by rushes and with a stand of Water plantain (*Alisma plantago-aquatica*).

Plate 4 The Great Silver water beetle
(Hydrophilus piceus)

(*a*) Egg cocoon (20 mm long). Note the fragments of *Hottonia palustris* attached to the cocoon and the 'mast'.

(*b*) The Great silver water beetle (*Hydrophilus piceus*) respiring at the surface. The antenna, covered with hydrofuge hairs, is curved round the anterior spiracle and breaks the surface film. Air is then taken in and stored beneath hairs on the ventral surface which gives the beetle a silvery appearance.

temperature on lotic species has not, so far, been accurately assessed, but from these and other experiments it must be inferred that species which have no means of creating a current of water over their bodies, must rely on a natural flow of water and that the optimum rate of flow will depend upon the temperature and oxygen content.

Habits and dispersal

The success of a species in establishing a breeding population depends not only upon its ability to cope physiologically with the various environmental conditions, but to a great extent upon its breeding habits and means of dispersal. As we have seen, temperature is of importance in deciding the onset of ovipositing in *Simulium*. But the survival of the young stages with little power of dispersal, in the presence of innumerable enemies, is dependent upon raising a first generation fairly rapidly and before the natural predators are present in large numbers.

Rare species, struggling to maintain a foothold in a habitat, offer opportunities for studying how their behaviour and breeding habits favour their continued existence.

The primitive crustacean, *Triops (Apus) cancriformis*, already mentioned, inhabits temporary pools and cannot withstand competition from the more highly evolved crustaceans and the insects inhabiting more permanent stretches of water. In order to survive long periods of drought when the temporary water in which it lives dries up, *Triops* produces latent eggs, capable of developing many months and sometimes even years, after they are laid.

The Great silver water beetle (*Hydrophilus piceus*) (Plate 4) the largest British water beetle, is restricted to a few places in the southern part of the country, Sedgemoor (Somerset) being one area in which it is still fairly common at certain times of the year.

The adult beetles first appear during April and temperature, as well as the growth of water weed, is probably important in regulating the onset of pairing and ovipositing. In the numerous peaty ditches or rhines transecting Sedgemoor, there are many which are not used by *H. piceus*. They tend to select those which

C

have a fairly thick covering of weed and especially where there is
a good growth of the Water violet (*Hottonia palustris*) (Plate 5).
This is the weed most enjoyed as food by the vegetarian adults

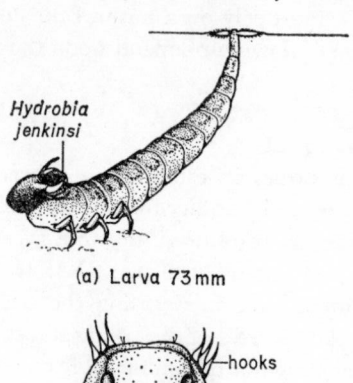

(a) Larva 73 mm

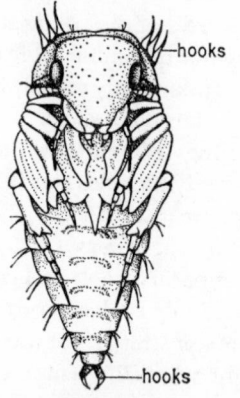

(b) Pupa 30 mm

Great silver water beetle
Hydrophilus piceus

Fig. 23. (a) Fully grown larva in the breathing and feeding attitude. The
head is thrown back so that the prey (in this case *Hydrobia jenkinsi*) is held
by the jaws against the dorsal part of the thorax. (b) Ventral view of pupa.
The three sets of hooks support the pupa so that it is surrounded by air
on all sides.

and the female also uses small fragments of the leaves, broken off
by the mandibles, to construct the large egg cocoon (Plate 6).
At about 15°C the forty-eight or so eggs, packed tightly within

the cocoon, take eleven days to hatch. The small carnivorous larvae emerge through a hole which they make at the base of the 'mast' of the cocoon. Equipped with strong mandibles (figure 23 (a)) they swim actively about and soon become dispersed among the thick weeds and commence to feed at once on small molluscs, to which their diet is entirely restricted. At first they will take only small species such as Jenkin's spire shell (*Hydrobia jenkinsi*). If food is plentiful growth will be rapid and at five or six weeks old they are capable of eating anything up to fifty or sixty of these small molluscs daily. At this stage they will tackle larger species

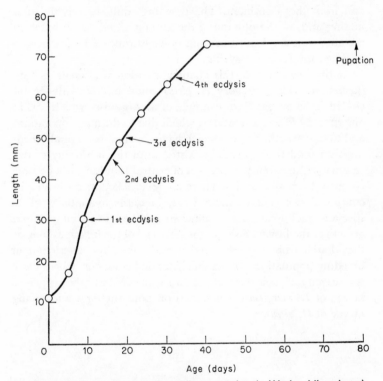

Fig. 24. Growth of larva of the Great silver water beetle (*Hydrophilus piceus*) June to August, at pond temperature.

such as *Pisidium* and various small planorbids, biting through the
shell to reach the soft tissues. When fully grown the larva is about
73 mm long (figure 24) and starts to search for a soft area in the
peaty walls of the rhines where it constructs a cell and pupates
(figure 23(b)). The spines keep the pupa braced within the cell so
that it is surrounded by air and can continue to breathe. After
six weeks or so, depending on temperature and humidity, the
adult beetle emerges and what happens next is a matter for
speculation. Balfour-Browne (1958) mentions records of the adults
being taken in light traps in Somerset and near the south coast.
This indicates that the beetles fly at night and possibly migrate to
and from the Continent. The breeding adults usually die after
mating and, in the opinion of the author, those which hatch in
the later summer, leave this country for European sites, returning
to breed the following spring.

At the beginning of this chapter mention was made of rare
species, in which category *H. piceus* could undoubtedly be included. It is probably an example of an organism which was at
one time more widespread but which due to drainage operations
and to the growth of towns which have covered its original breeding sites (and these certainly existed until the beginning of this
century within twenty kilometres of central London), has become
confined to small colonies. There are several factors governing its
continued existence in these areas. First the availability of its
specialist molluscan diet in sufficient numbers to maintain rapid
growth of the larvae. Second, the dispersal of the larvae as soon as
they hatch from the egg cocoon and third, the restriction of
breeding populations to certain rhines on Sedgmoor, where there
is a scarcity of their natural predators, particularly the voracious
larvae of *D. marginalis* which feed on both the eggs and young
larvae of *H. piceus*.

5

Studies of spatial distribution and density (1) In still water

Any community of animals and plants functioning together with their non-living environment can be called an *ecosystem*. In fact ecosystems exist everywhere—in a wood, a field, the sea, or even a crack in a wall supporting a community of animals and plants.

Fresh water offers some of the best examples of ecosystems. A lake or pond, a river or stream, are all ecosystems. Yet each is really a complex of systems. Each physical area of a pond, such as the surface film, in the weedy area or at the bottom in the mud, is a separate system. Streams and rivers also have their complex of ecosystems.

We talk of ecosystems often without much attempt at defining boundaries, either physical or physiological, but we can apply more precise methods of description to microhabitats which exist within an ecosystem. A stone lying in the bed of a stream can form a microhabitat for *lithophilous* organisms, that is those animals and plants using stones as shelter and which live together in a small community of their own. The roots of a water plant such as Water plantain (*Alisma plantago-aquatica*) may support a number of organisms, likewise there may be small animals inhabiting its hollow stems. Each is a microhabitat.

The ecological niche

The ecological *niche* of a species is a term often used by ecologists to describe the role that an organism plays within an ecosystem. As Odum (1963) puts it: 'The habitat is the "address", so to speak, and the niche is the "profession".' A single species may occupy different niches within different ecosystems. For instance, the Water louse (*Asellus aquaticus*) in the context of open water, forages about among pondweeds feeding upon epiphytic algae, while the same species, crawling on the surface of the ooze, feeds on organic debris.

Freshwater plants, too, have their niches but because they are not mobile, they tend to occupy a narrower niche than aquatic animals. The water lily will always be found rooted in the mud of shallow, still water where its floating leaves can reach the surface. The various species of duckweed also live in still water where their roots can hang down from the plants floating on the surface. The larger emergent plants, such as the Arrowhead (*Sagittaria sagittifolia*) and *A. plantago-aquatica*, colonize shallow water at the edge of ditches, canals, or ponds where they can root in the mud with their leaves growing up above the surface of the water (Plate 3 (b)).

Preliminary surveys

In order to become familiar with a locality, to discover what problems it presents and which are worth investigating in greater depth, it is important to begin by making a preliminary survey of the area selected for study. The rest of this chapter describes how such a survey could be carried out.

A small stretch of the now disused Tiverton Canal offers a possible site on which a survey can be made. There is a distinct zonation of plants growing in the silted up waterway, where flow is negligible. The banks are broken in many places so that they slope down to the water from the towpath, offering quite a wide area for the colonization of marsh plants, while in others the sides are nearly vertical.

Vegetation maps

A vegetation map should be made in order to show the main distribution of plants and their relative abundance within a specified area. Such a map can also have importance because of the direct association of the animal communities living in or among the vegetation. An enormous amount of time can be wasted, however, in making a detailed plan of an area when, for a preliminary survey, a record of the main distribution of the plants is all that is required.

Figure 25(a) shows a map, made in late June, of a short stretch of the Tiverton canal (Plate 6 (*a*)). A length of plastic clothes line, marked at 1 m intervals with coloured Sellotape, was laid along the towpath and from this two similarly graduated lines were stretched at right angles across the canal 1 m apart. The positions of the chief communities of plants within this belt transect, were plotted. By moving one transverse line another metre along the horizontal, the next belt could be mapped, and so on.

A line stretched between two poles at A and B and levelled by means of a spirit level, formed a datum line. From this, vertical measurements at 25 cm intervals, of the height of the ground below datum were taken and the profile was drawn (figure 25(b)).

Zonations and successions in plant communities are often best studied along lines perpendicular to the zones. These are called *transects* of which there are various kinds. One of the most useful is the ladder transect which shows how changes in individual species and community structure are associated with the variations in the habitat. In order to construct a ladder transect it is necessary to assess the proportion of an area covered by a given plant and this may give different results from that obtained by recording the number of plants. From the ecological point of view the former is preferable and the more important for a general assessment of the community. A cover/abundance scale is given below:

1 = Scant—covering less than 1/20 of the ground or water surface

2 = Covering 1/20 to 1/4 of the ground or water surface

3 = Covering 1/4 to 1/2 of the ground or water surface

4 = Covering 1/2 to 3/4 of the ground or water surface

5 = Covering 3/4 to all the ground or water surface

Using such a scale is often difficult in practice and another method is to group the plants into categories as to whether they are abundant, frequent, occasional, or rare. Although this gives

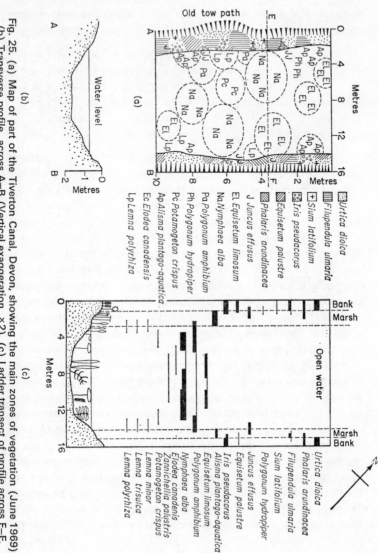

(b)

(a)

Metres

Old tow path

Water level

B

A

Metres

	Urtica dioica
	Filipendula ulmaria
+ Sium latifolium	
	Iris pseudacorus
	Equisetum palustre

Ph Phalaris arundinacea
J Juncus effusus
El Equisetum limosum
Na Nymphaea alba
Pa Polygonum amphibium
Ph Polygonum hydropiper
Pc Potamogeton crispus
Ap Alisma plantago-aquatica
Ec Elodea canadensis
Lp Lemna polyrhiza

(c)

Metres

Open water

Bank
Marsh

Marsh
Bank

Urtica dioica
Phalaris arundinacea
Filipendula ulmaria
Sium latifolium
Polygonum hydropiper
Juncus effusus
Equisetum palustre
Iris pseudacorus
Equisetum limosum
Alisma plantago-aquatica
Equisetum limosum
Polygonum amphibium
Nymphaea alba
Elodea canadensis
Zannichellia palustris
Potamogeton crispus
Lemna minor
Lemna trisulca
Lemna polyrhiza

Fig. 25. (a) Map of part of the Tiverton Canal, Devon, showing the main zones of vegetation (June 1969).
(b) Transverse profile across A–B (Vertical exaggeration ×2). (c) Ladder transect of profile across E–F.

a satisfactory general impression of an area it does not take into account the size of the plants tending to make small plants such as the duckweeds assume more importance than the larger ones. Probably a compromise of the two methods is the best to aim at in making a preliminary survey of this kind.

A ladder transect (figure 25(c)) was constructed along the line E–F, marked in figure 25(a). The profile at C–D was drawn at the bottom, from measurements made using a datum line as for the profile A–B, and a list of the chief species of plants was made, in an order which conveyed the changes in the vegetation along this transect. The frequency variations for each species were plotted as histograms horizontally along lines above the appropriate part of the profile. The thickness of the histogram lines also, in this case, takes into account the amount of cover, and vertical dotted lines indicate the boundaries of the main plant zones.

If seasonal changes are to be recorded over a period of time, temperature, light intensity, pH, and the rate of flow of water, can be plotted beneath the diagram.

A series of such transects made across the area can show how changes in communities and in individual species are associated with changes in the configuration of the banks or depth of the water at different points.

Distribution of the animals

Within the larger area of the canal there were, of course, a number of microhabitats, each with their own populations of animals and plants. Table 1 lists some of the animal species that were found in five microhabitats in different parts of the area surveyed.

Obtaining specimens from these microhabitats involves a different method of collection for each of the five areas. It will also be evident that some are much smaller and more restricted than others and this brings us back to the point made at the beginning of this chapter concerning the difficulty of defining the boundaries, both spatial and physiological, of a microhabitat.

In terms of this survey, the microhabitat of 'open water in areas of weed' is clearly one which is ill-defined. For a preliminary and

Table 1

The distribution of animal species in five microhabitats in the area surveyed of the Tiverton Canal in Devon.

	Surface of water	Open water in areas of weed	On upper or lower leaf surfaces of *N. alba*	Inside stems of *A. plantago-aquatica*	Mud round roots of *A. plantago-aquatica*
COELENTERATES					
Chlorohydra viridissima			√		
PLATYHELMINTHES					
Dendrocoelum lacteum			√		√
Polycelis nigra			√		√
ANNELIDS					
Eiseniella sp.					√
Erpobdella octoculata			√		√
NEMATODA					
Dorylaimus stagnalis				√	√
CRUSTACEANS					
Asellus aquaticus					√
Gammarus pulex					√
INSECTA					
Hyphydrus ovatus		√			
Corixa sp.		√			
Ilyocoriscimicoides		√			
Coenagrion puella		√(nymphs)	√(nymphs and adults)		
Gyrinus natator	√				
Gerris sp.	√				
Hydroporus pictus		√			
Notonectaglauca		√(nymphs and adults)			
Triaenodes bicolor		√(larvae)	√(eggs and larvae)		
Chrionomus sp.				√	√
ARACHNIDA					
Hydrarachna sp.		√			
MOLLUSCA					
Ancylus lacustris			√		√
Sphaerium sp.					√
Pisidium sp.					√
Limnaea pereger		√			
Planorbis corneus		√	√		
P. planorbis			√		

less detailed survey, the method of collecting the organisms in this area can only be by using a net to sweep the water among the weeds. Some of the organisms caught in the net will be those swimming about in the water, others may well be those which were resting on the vegetation or which were actually attached, in some way, to the water plants. Only by adopting more careful methods could one subdivide this habitat into two more clearly defined microhabitats—that of open water between the weeds and that of the organisms found adhering to the weeds themselves. Likewise, collecting from the area of mud around the roots of *A. plantago-aquatica* involves the rather haphazard method of pulling up the plant and washing the roots free of mud in order to collect the organisms therein. For more detailed work, more methods would have to be employed, involving the separation of those animals found in the mud associated with the roots and those actually attached to the roots and burrowing into them.

Indications for more detailed work

These rather rough and ready methods can give some idea of the various species present, and of their distribution. It can also indicate where more information would be useful. Is the Freshwater limpet (*Ancylus lacustris*), for instance, confined to the upper or undersides of the floating leaves of the White water lily (*Nymphaea alba*), and do they occur on the leaves of other plants? Are they distributed according to food preferences, to light intensity, or to other factors? Exposure to wind as well as their reproductive activities may also play a part.

The preliminary survey indicated that in nearly all cases, *A. lacustris* was found adhering to the undersides of the leaves of *N. alba*. More detailed counts showed that the average of seven samples was 37 per 10 cm² of leaf surface, whereas the average of seven counts for the same species occupying the undersurface of the leaves of Amphibious persicaria (*Polygonum amphibium*) was 7 per 10 cm². These results might be worthy of further investigation in order to discover whether there was a distinct preference for the leaves of *N. alba*, and if so, the reasons for this. One reason might be that the epibionts of *N. alba* were more preferable as food than those of *P. amphibium*.

At the time of year when the survey was made, many species such as the Damselfly (*Coenagrion puella*) and the Caddisfly (*Trianodes bicolor*) were using the stems and leaves of *A. plantago-aquatica* and of *N. alba* for ovipositing. This could well be a seasonal cause of 'clumping' of a species rather than of the more even distribution of that species at another time of the year. Adults of the red-coloured water mite (*Hydrarachna* sp.) (figure 26 (a)) were

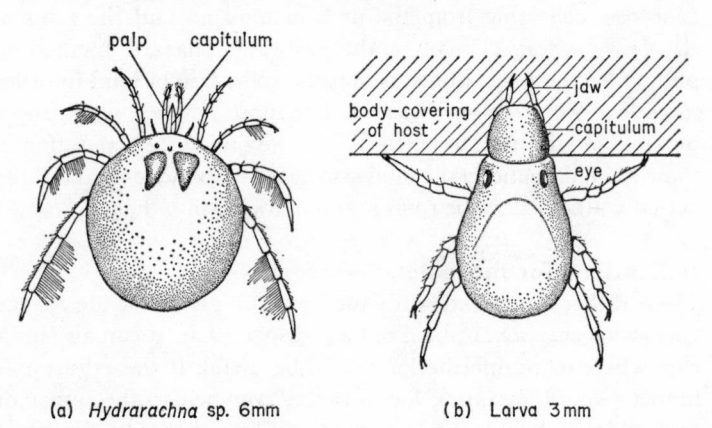

(a) *Hydrarachna* sp. 6mm   (b) Larva 3mm

Fig. 26. (a) One of the largest water mites. The adults of this genus are free-living and carnivorous. (b) The larva is parasitic and has only three pairs of legs. There are strong jaws in front of the false head or capitulum which is embedded in the tissues of the host.

much in evidence swimming about near the surface, probably searching for suitable leaf-surfaces in which to pierce holes for their eggs. Although only a few larvae of this mite (figure 26 (b)) were caught in net sweeps in open water, they were probably there in numbers not as free-living individuals, but as ecto-parasites of insects. It is by no means unusual to find adults of the Great diving beetle (*Dytiscus marginalis*) or of the Water scorpion (*Nepa cinerea*) parasitized by two or three of these larvae which have prominent false heads, really enlargements of the forepart of the body, embedded in the soft tissues of the host. The complicated life-history of such a parasitic species is one of the reasons for its distribution and for its occurrence in different microhabitats at different stages in its life-history.

From this preliminary survey it becomes obvious that besides the physical factors of the environment which control the distribution of an animal, there are also biotic factors operating such as competition between species, predation, behaviour or methods of concealment, all of which may be of importance in the spatial and seasonal distribution of members of the community. The significant point to remember is that each organism is distributed according to its own particular links with the environment and that where several life stages are involved, each may present a different distribution problem.

Estimating populations

Investigation of the reasons for the distribution of an organism or of a group of organisms will lead to the question of how one estimates the size of a population by taking samples. The object of taking samples is to measure in a small population (the sample) characteristics of the larger population from which the sample was taken. It is of great importance, therefore, to select a site where a reasonable sampling programme can be carried out.

Estimating the size of a population of small animals which are moving about is obviously harder than making the same kind of investigation of a plant population. Nevertheless, many communities of animals in still water are at least partially isolated and an estimate of numbers, sufficiently accurate for the purpose of comparing one site with another, can be made. This is true of the surface dwellers such as pond skaters or water measurers which never dive beneath the surface. These are hemipterous insects which prey on other small insects falling on to the water surface. It will probably be evident that they congregate in different regions of a pond, perhaps beneath trees from which small aerial insects are likely to drop, or in open water where there is a hatch of gnats taking place.

The preliminary survey will have indicated very roughly whether the animals are aggregated in certain areas and sparse in others, also whether the population is composed of mature or immature stages or a mixture of both. It may be that only a rough estimate of numbers is required for comparing one population with another. But if a sampling programme is called for which requires accurate results and involves a number of species with a wide range of densities, we at once enter the realm of the statistical ecologist which is outside the scope of this book. In this case reference should be made to the books or papers mentioned in the list of references at the end of the chapter.

For animals sparsely distributed or which are very active, the Lincoln Index on marking and recapture method can be used. This involves the marking of a certain number of individuals and returning them to the water. The method is only possible for those species which can be marked in some fairly permanent way. Fish can be marked with a small metal tag attached to the gill cover or to the tail. Spots of cellulose paint or nail varnish can be painted on the shells of molluscs and even to mark insects.

Assuming that the marked individuals mix at random and that a fairly large population of marked individuals can be built up quickly, the following formula can be used to calculate the whole population, X:

$$X = \frac{ab}{r}$$ where a = number of individuals marked and released

b = total number of individuals recaptured
r = number of recaptured marked individuals.

If a series of capture and release operations are carried out it is possible to follow fluctuations in a population and to gain some idea of the factors such as births, deaths, immigration, and emigration which can alter its size. However, there can be many sources of error and for more accurate field studies the paper by Bailey (1952) should be consulted.

6

Studies of spatial distribution and density (2) In running water

Many streams and rivers, even over a comparatively short stretch of water, show variation in width from bank to bank, in depth, and in the speed of the current. But from the point of view of the stream ecologist, it is the nature of the stream bed which is one of the most important factors determining the character of the stream as a whole; that is, whether the bottom is generally firm or whether it is composed of soft, shifting mud and sand. In many cases, the two types of stream bed can be found alternating in the same stream.

In parts where the basin is composed of rocks or boulders, which do not normally shift except under extreme cases of flooding, there may be deep pools resembling ponds, as well as hard-bottomed rapids supporting a limited flora and fauna which are specially adapted to living in swift currents or whirlpools.

Where the stream or river flows less rapidly, sediments can be deposited in deep pools in which a considerable development of phytoplankton can occur.

Another important factor affecting the nature of different parts of a stream is the degree of illumination. Where a stream flows through meadowland, plants growing on the bank and in the water receive full light and will flourish bringing with them a richer fauna. But in places where light is cut off by overhanging trees, fewer plants will be found. Leaves falling into the water will enhance the richness of the mud which will support a different fauna.

The regions of a stream
At first sight many streams, particularly in areas where the water flows fairly swiftly, might appear to be fairly barren of life.

Turn a stone, however, and numerous small animals can be seen scuttling about on its surface, or shake a handful of moss underwater and many other small creatures will be washed out. Diatoms and other unicellular algae are attached to the stones and on these feed the larger organisms such as ephermerid nymphs and molluscs.

Swift-flowing head-streams, characterized by a small temperature range, shallowness, and infrequency of vegetation, present their own problems. Those organisms which have become successful colonists, although few in terms of species, are usually plentiful in actual numbers. The majority are *microphages*—that is those which feed upon microscopic particles of food, straining the water for detritus and plankton, borne along in the current.

Besides those which take shelter in moss or beneath stones, there is a group of organisms deserving special attention: small animals which live in the thin film of water surrounding the surface of partially exposed stones or weed. These constitute what are sometimes called the *hygropetrical* fauna. The larva of the midge (*Dixa maculata*) is an example (figure 27(a)). Living in the thin film the larva wriggles, bent in a typical U-shape, against the stone and thrusts its hind end, bearing the spiracles, through the surface of the film to obtain a supply of atmospheric air.

Lower down, the head-streams join to form a wider watercourse with a larger volume and a swifter current. The constant erosion of the stream bed causes an increase in the amount of particles carried downstream, but the current is usually too great for the development of plankton. The current and the force of the water makes these regions intolerable for any organisms such as dipteran larvae, which need to take air at the surface. Many are *lithophilous* forms—that is those living on the surface of stones, either above or beneath them.

The microhabitats of a stream

The stretch of the stream shown in Plate 7(*a*) is typical in possessing a variety of microhabitats. *Cladophora* sp. and the Water moss (*Fontinalis antipyretica*) are the only weeds which have become

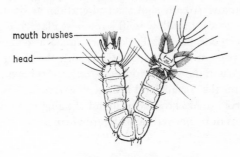

(a) Larva of *Dixa maculata* 5mm

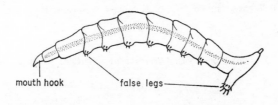

(b) Larva of *Limnophora* sp. 8mm

Fig. 27. Two dipteran larvae common in running water. (a) *Dixa maculata*, a hygropetrical species that feeds on stream detritus collected by whisking the mouth brushes to and fro. (b) *Limnophora*, often found attached to weed in the strong current of a waterfall, by means of its mouth hooks and false legs.

established on the weir itself. The larva of the dipteran, *Limnophora* sp. (figure 27(b)) is able to cling to the weed by means of mouth hooks and the leeches *Erpobdella octoculata* and *Glossiphonia complanata* by their anterior and posterior suckers. Large colonies of chironomid larvae (figure 14) also manage to live in the deeper layers of the thick bed of *Cladophora*.

Immediately below the weir there is a small whirlpool of deeper water, where plant life is almost absent. The Stone loach (*Nemacheilus barbatula*) (figure 28(a)) with its dorso-ventrally flattened body, hides among the stones where young eels also abound.

Lower down the stream shows an alteration in depth from 30 cm or more to very shallow swift stretches (Plate 7 (*b*)).

In the deeper pools forming backwaters, fauna associated with still water are present. Pondskaters (*Gerris* sp.), Whirligig beetles (*Gyrinus natator*), and the Water cricket (*Velia currens*) hawk smaller insects falling on the surface and even a slow-moving insect like the Water scorpion (*Nepa cinerea*) is found clinging to submerged plants by means of its hooked tarsi (figure 28(b)).

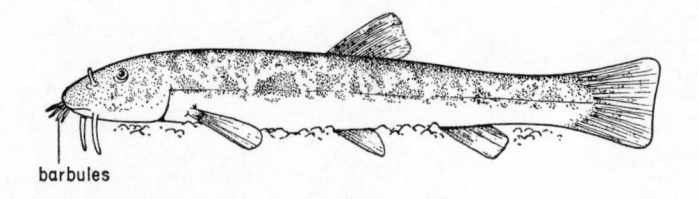

barbules

(a) Stone loach (*Nemacheilus barbatula*) 110 mm

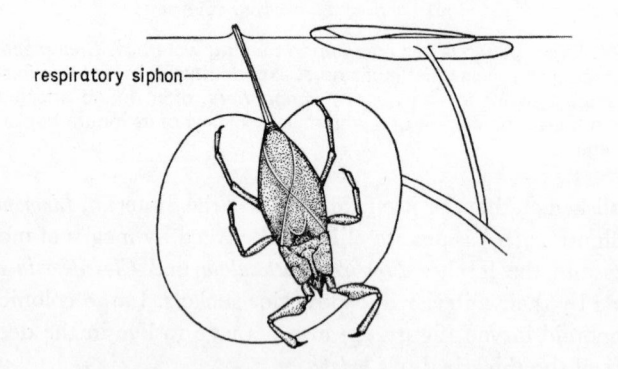

respiratory siphon

(b) Water scorpion (*Nepa cinerea*) 32 mm

Fig. 28. (a) The dorso-ventrally flattened body of the Stone loach and the barbules on its head, adapt it to living on the stream bed. (b) The sluggish Water scorpion manages to maintain station by clinging to submerged weeds. Air is taken in at the surface by means of the long respiratory siphon at the end of the abdomen.

The rapid, shallower parts of the stream support a lithophilous fauna adapted to withstand the current in various ways. Two species of ephemerid nymphs are particularly numerous: *Baetis* sp. which has a delicate spindle-shaped body, manages to cling to the stones hitching on by its curved claws (figure 29 (a)), and the more robust nymphs of *Ecdyonurus venosus* (figure 29(b)) in which the body is flattened dorso-ventrally. The legs have expanded coxae and bear claws which also assist in maintaining contact with the surface of the stones.

Two species of caddis larvae are also typical of the swifter reaches, as they are in many streams with stony bottoms. *Hydropsyche angustipennis* (figure 29 (c)) builds a funnel-shaped silk net, closed at one end, with a wide mouth facing upstream. Floating matter of both animal and plant origin, is caught in the net and seized by the larva as food. The net sieves the food and long bristles at the end of the abdomen, keep the net clean. *Rhyacophila dorsalis* (figure 29 (d)) builds no larval case but lives free, creeping about beneath stones in search of chironomid larvae and other small animals. It uses the long anal appendages for gripping the substratum. Both species pupate within cases constructed from small pebbles, which are glued to the surface of larger stones.

The larvae and pupae of *Simulium* sp. (figure 29 (e)) colonize the stones of the stream bed in large numbers. The larvae are attached by a special tail organ while the mouth brushes, facing upstream, constantly sieve the current for particles of food. The fibrous cocoon of the pupa is attached by threads, the mouth of the cocoon always facing upstream. The position of a dense colony of *Simulium* larvae on the most exposed part of a stone, might be attributed to the selection of current of a certain speed. This distribution could also be the result of predation by ephemerid or plecopteran nymphs, which leave only those where the current has been too swift for them.

Ambühl (1959) found that the Freshwater shrimp (*Gammarus pulex*) could not withstand a current of more than 45 cm/sec. Although its body is laterally flattened and it can actively seek shelter by swimming, its powers of locomotion are not great and

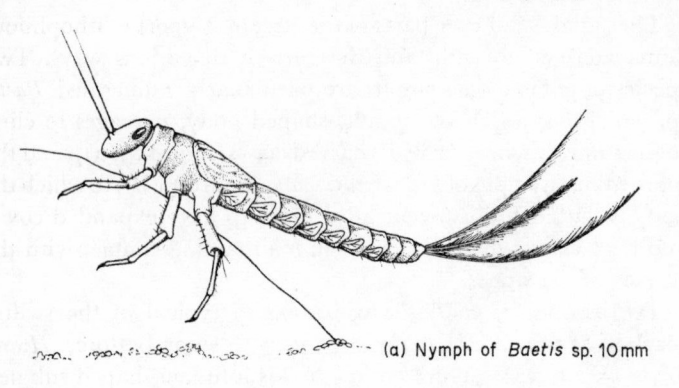

(a) Nymph of *Baetis* sp. 10 mm

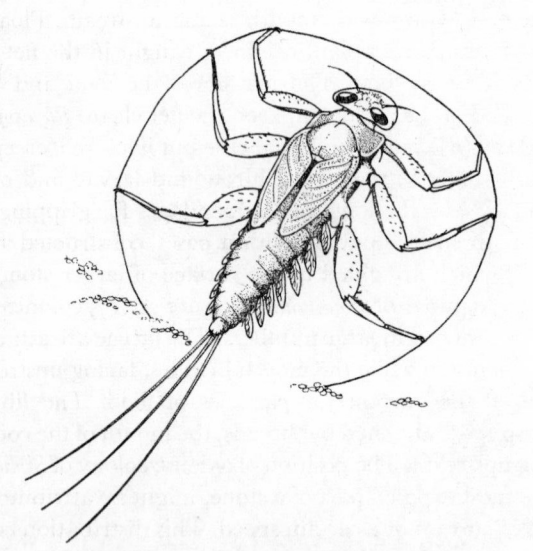

(b) Nymph of *Ecdyonurus venosus* 40 mm

Fig. 29. Some lithophilous insects typical of fast-flowing streams. (a) and (b) Two mayfly nymphs. Both cling to stones by means of strong hooks on their legs. (c) and (d) Two caddis larvae. *Hydropsyche* spins a net and a loose shelter of silk to which stones are often attached. *Rhyacophila* lives free and does not construct a case. (e) A group of *Simulium* larvae. The pupa breathes by means of two tufts of tracheal filaments which project from the fibrous cocoon attached firmly to a stone.

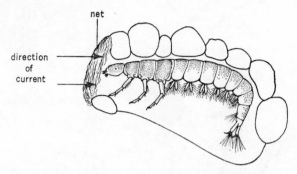

(c) Larva of *Hydropsyche angustipennis* 20 mm

(d) Larva of *Rhyacophila dorsalis* 20 mm

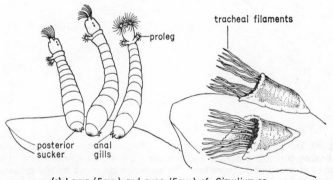

(e) Larva (5 mm) and pupa (5 mm) of *Simulium* sp.

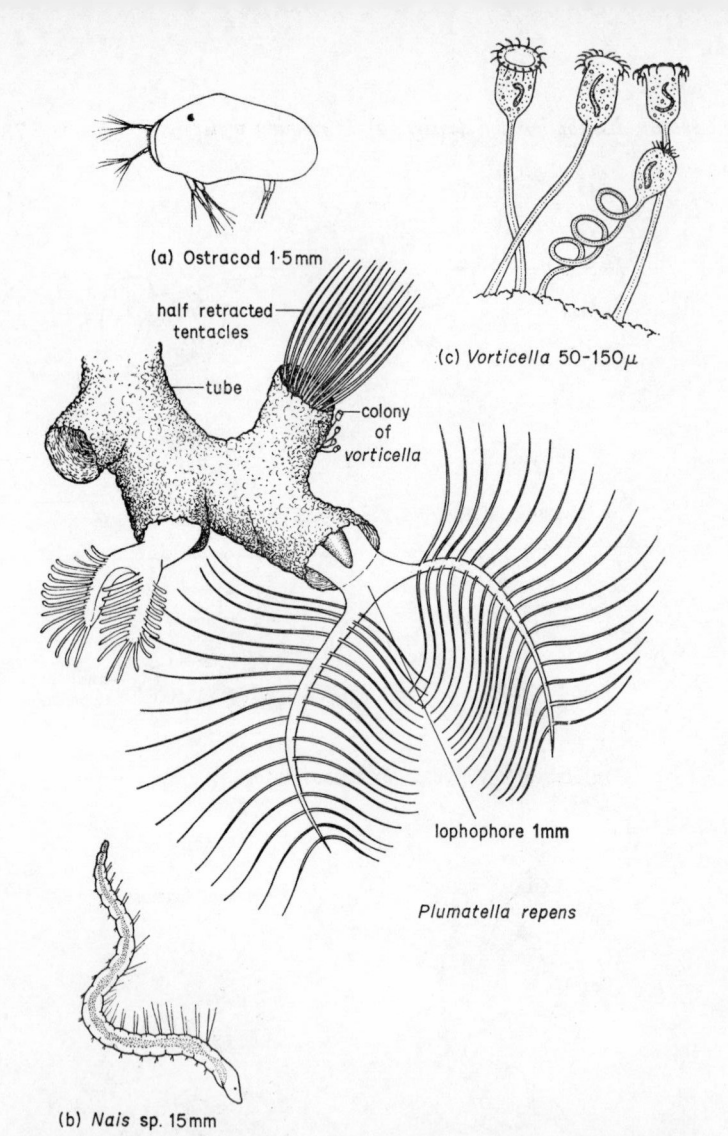

(a) Ostracod 1·5mm

half retracted tentacles

tube

colony of vorticella

(c) Vorticella 50-150μ

lophophore 1mm

Plumatella repens

(b) Nais sp. 15mm

Fig. 30. The polyzoan, *Plumatella repens*. Part of a colony found adhering to broken concrete blocks in Exton stream, Devon. Colonies can cover an area of as much as a square foot. The tentacles are borne on a horseshoe-shaped lophophore which is retractile. Some of the organisms found sheltering in the dense colonies are (a) ostracods and (b) *Nais*, a small annelid. (c) Colonies of the bell animalcule, *Vorticella*, attached to the tubes of *Plumatella*.

it has developed no special structures for clinging to stones. Nevertheless, it is found in the less swift parts, its numbers probably kept low by predatory fish such as the Miller's thumb (*Cottus gobio*) and the eel (*Anguilla vulgaris*).

One of the moss animalcules (Polyzoa), *Plumatella repens*, is a sessile organism occurring in the stream on the broken concrete blocks which were washed into the stream bed during the severe floods of 1960. These offer the rough surface needed for the development of the colonies. *Plumatella* affords a microhabitat of its own for several species often found within a colony. Rotifers, 'water bears' (Tardigrades) copepods (figure 30) and even larger crustaceans such as *Asellus* and *Gammarus* may shelter among the branches of the colony.

Probably one of the most successful animals of the swifter reaches of the stream, is the River limpet (*Ancylastrum fluviatile*) which clings to stones by clamping its shell firmly to the surface. The edge of the shell is soft and so fits any irregularities (figure 31).

The influence of current
Water movement, because of its affect upon the substratum, is of prime importance in the distribution of species and in deter-

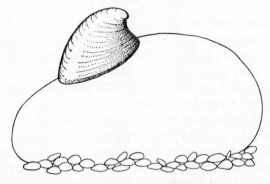

The River Limpet (*Ancylastrum fluviatile*)

5 mm

Fig. 31. The edge of the shell of the River limpet (*Ancylastrum fluviatile*) is soft and fits irregularities of the stone to which it clings.

mining the size of populations. Microhabitats abound and although these are usually associated with current, in backwaters they can resemble conditions in still water. The presence of a boulder or even a small stone can alter the speed and direction of the current and can also create small eddies (figure 32). As the water strikes the stone, the current is diminished and causes water in contact with the stone to flow more slowly over it. On the downstream side there may be a back flow in the immediate area of the stone, causing an eddy and the scouring out of the stream bed. In this way several microhabitats are created in the vicinity

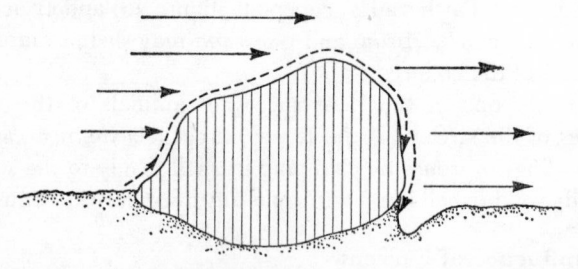

Fig. 32. The effect of current flowing over a stone in creating a microhabitat. Firm arrows show the general direction of the current, broken arrows the direction of the current flowing over the stone.

of the stone. The current will be greatest on the upstream side diminishing on the downstream side and causing the water to be thrown back on to the stone on that side. Areas beneath the stone on the downstream side will be more sheltered and if other factors than current, such as predation and food, could be eliminated, there might well be found a distribution of lithophilous species according to their individual tolerance of current speed.

Ambühl (1959) found that certain species occurred over a wide range of current speed. His graph (figure 33) shows that *Hydropsyche* spp. are found in greatest numbers at 60 cm/sec, while at the other end of the series, *Gammarus pulex* reached its peak numbers at 15 cm/sec. These figures are, however, open to criticism for several reasons. For one thing the current speed was recorded not where the organisms were found but above the bottom. Also,

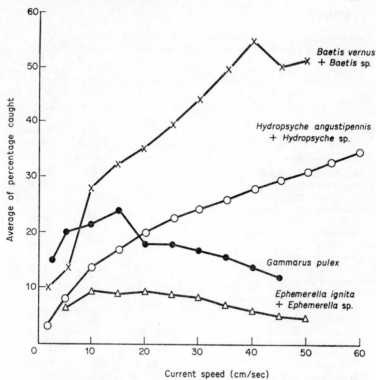

Fig. 33. Average of percentage of species caught at various current speeds. (After Ambühl, 1959.)

apart from *G. pulex*, each curve represents a mixture of species, the figures for each curve being calculated as a percentage of the whole catch of that group, and the totals varying for each. This illustrates how faulty methods of sampling can give rise to a misleading interpretation of results.

Dorier and Vaillant (1954) showed the relationship between current speed and the occurrence of other species (Table 2). To withstand a current speed up to 240 cm/sec an organism must have very special means of attachment to the substratum. The leech, *Glossiphonia complanata*, possesses a sucker at the anterior and

Table 2

The relation between current speed and the occurrence of certain species. Figures are current speed in cm/sec (selected from Dorier and Vaillant, 1954).

Species	Speeds at which the species was found in nature		Max against which species will ascend	Speed at which washed away
	min	max		
Agrion sp. (Odonata)		10	54	77
Polycelis felina (Turbellaria)		10	44	99
Dendrocoelum lacteum (Turbellaria)		10	37	76
Glossiphonia complanata (Hirudinea)		10	37	240
Planaria alpina (Turbellaria)	10	14	140	143
Limnaea pereger (Mollusca)	10	14	117	202
Ancylastrum fluviatile (Mollusca)	10	24	109	240
Heptagenia lateralis (Ephemeroptera)		28	140	188
Gammarus pulex (Crustacea)	10	40	44	99
Theodoxus fluviatilis (Mollusca)	10	78	109	240
Simulium ornatum (Diptera)	14	114	117	240
Rhyocaphila sp. (Trichoptera)		125	100	200

posterior end of the body which is flat and offers little resistance (figure 34(a)). *G. complanata* produces a gelatinous cocoon which is held beneath the parent's body. When the young hatch they attach themselves to the ventral surface of the parent where they remain until they have digested their reserves of yolk and can feed independently. *Erpobdella octoculata* on the other hand, lays a cocoon which is at first lemon-shaped but which the leech flattens so that it lies snugly attached to the substratum where it hardens and becomes dark brown (figure 34 (b)).

The gastropod molluscs, *A. fluviatile* and *Theodoxus fluviatilis* are both capable of clamping their shells, by means of a foot, to the rock surface. Indeed *A. fluviatilis* is found on top of stones in a stream bed as often as it is found beneath them, where the current is likely to be less. However, Dorier and Vaillant's observations are subject to various interpretations. Records of the distribution of

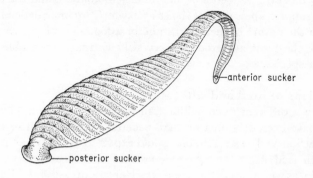

(a) *Glossiphonia complanata* 15–45 mm

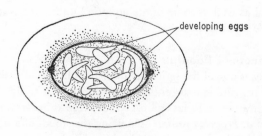

(b) Egg cocoon of *Erpobdella octoculata* 2 mm

Fig. 34. (a) The leech, *Glossiphonia complanta*, can withstand current speeds of up to 240 cm/sec without being washed away by holding on to the substratum with its strong suckers. (b) Egg cocoon of *Erpobdella octoculata*.

organisms relative to the current, made in one stream, might not
hold for another where the substratum is different and where the
availability of food is not the same.

As we indicated in Chapter 4, the effect of current on the distribution of organisms must be assessed along with other factors.
The faster a stream flows, the more particles pass a given point
in a given time. Those animals such as *Hydropsyche* which exploit
these particles as food, must endeavour to make a compromise
between current speed and their ability to hold on, and the best
food supply. Those such as *A. fluviatile*, which are more dependent
upon the nature of the substratum for a foothold, may turn out
to be the most numerous in very swift currents, untenable by
other species.

Regions of mud and silt

Quite a different microhabitat can be formed where mud and
silt have been deposited in small backwaters. These will be rich
in detritus and as a result, one would expect to find those animals
which feed upon the richer deposits of animal and vegetable
origin, as well as upon the weeds which are more easily established
in these regions. The larvae of different species of flies such as the
cranefly (*Tipula* sp.), the midges (*Tanypus* sp. and *Corynoneura* sp.)
and the rat-tailed maggot (*Tubifera* sp.) (figure 35) are found in
the mud as well as some of the aquatic lumbricid worms such as
Eiseniella tetrahedra.

The effects of flooding

Flooding is one of the serious climatic conditions affecting stream
organisms. Sudden torrential rain can cause a rapid rise in the
rate of flow beyond that against which some animals can maintain
a foothold. *Hydrobia jenkinsi* colonizing the stones of a small brook
in large numbers can be completely washed out of the stream by
heavy rain. This was the case in August 1968 when the total
population of this snail was washed out of a brook in Devon and
down to the sea. The same flood had little effect on the numbers of
A. fluviatile or of *E. octoculata*.

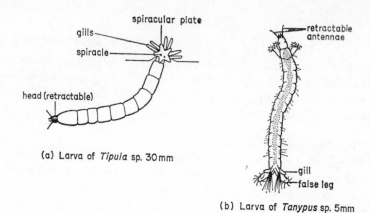

(a) Larva of *Tipula* sp. 30 mm

(b) Larva of *Tanypus* sp. 5 mm

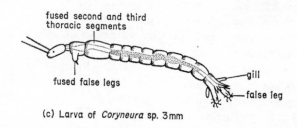

(c) Larva of *Coryneura* sp. 3 mm

(d) Larva of *Tubifera* sp. up to 55 mm

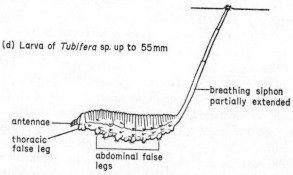

Fig. 35. Four genera of dipteran larvae commonly found in river mud and silt. *Tipula*, *Tanypus*, and *Coryneura* possess simple gills. *Tubifera* has a telescopic breathing tube which can be extended in deeper water so that it reaches the surface. When fully extended it is several times the length of the body. Probably all these larvae are capable of direct gaseous exchange through the thin body wall.

In the lower reaches of a river, plankton may be fairly abundant, but in a swift upland stream or after flooding, even a normally slow, meandering stream can be denuded of plankton. This, in turn, will impoverish the microphages.

The significance of invertebrate drift

Except after flooding, drift material is present in most rivers and streams and consists not only of vegetable matter but also of numerous invertebrates, chiefly aquatic in origin, although some terrestrial species falling into the water from overhanging vegetation, are often present in the surface drift.

Investigation of the drift material shows that its composition, in so far as its invertebrate content is concerned, varies in different parts of a stream and at different times of the day as well as at different seasons of the year.

In the same way as the small organisms living in the ooze of still water are of importance as food for the larger creatures, so the invertebrate drift of a stream is an important constituent of the food of larger stream animals, especially fish.

Elliott (1967) working on a Dartmoor upland stream, sampled invertebrate drift by means of a surface net made of nylon sifting cloth, with a rectangular mouth of known dimensions. This was attached by hooks to a metal frame at each end of which were floats (figure 36(a)). The whole apparatus was moored to each bank of the stream by means of ropes and secured so that the bottom of the frame was 7 cm below the surface. In this way the volume of water passing through the net over a certain period of time could be calculated. Records were taken each month when the nets were emptied every 3 h over a period of 24 h and the numbers of organisms entering during each 3 h period were noted.

In addition to the drift samplers placed at strategic points along the stream, several modified high-speed plankton samplers (figure 36(b)) were also used. These consisted of a metal tube enclosing a nylon net of 15·5 meshes/cm. This size of mesh was sufficiently small to trap organisms such as small insect larvae, 1–2 mm long, without becoming clogged. A flow-meter, fixed at the rear end of the tube recorded directly the volume of water passing through the net. These tubes were kept in place by iron rods attached to side brackets and driven into the stream bed.

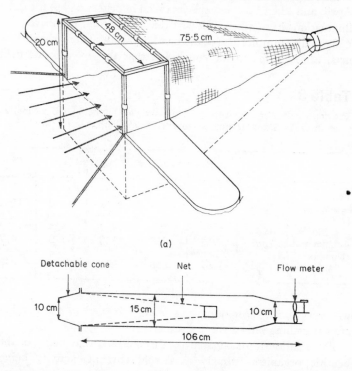

(a)

(b)

Fig. 36. (a) Surface net for sampling invertebrate drift. (After Elliott, 1967.) (b) Modified high-speed plankton sampler. (After Elliott, 1967.) The net is held in place by a detachable cone which reduces the effective sampling aperture to an area of 78·5 cm².

Elliott's results indicate several important points. The density of aquatic drift, measured for the months June to October in two successive years, appeared to be fairly constant for the same month in each year but not for different months (Table 3). Furthermore, at all seasons more aquatic invertebrates were found in drift samples taken at night than during the day, drift numbers probably being related to light intensity. Figure 37 shows results from drift samples, taken three hourly during 24 hours in March, April, and May. For most species taken in drift samples, the daily fluctuations in numbers followed that of the total numbers.

By taking samples also of the benthos, at sites below the surface nets, Elliott came to the conclusion that the invertebrate drift

Table 3

The density of aquatic drift sampled from July to October in two successive years in Walla Brook, Dartmoor (selected from Elliott, 1967).

		June	July	Aug.	Sept.	Oct.
Volume sampled (1000 1/24h)	1963	2144	1560	2768	1600	1592
	1964	1048	720	624	984	1572
Aquatic drift (Number/ 24h)	1963	763	1036	1998	770	477
	1964	267	619	261	217	443

was very local, most organisms being returned to the benthos after travelling only a short distance.

The relationship suggested by this study is that many of the benthic organisms display a diurnal rhythm, activity being strongest at night. *Baetis rhodani*, *Ecdyonurus forcipula*, *Heptagenia interpunctata* and other nymphs show a strong negative phototaxis as well as a strong thigmotaxis, activity reaching a maximum just after sunset, when their negative phototaxis no longer operates and they tend to move to the top of the stones where they can forage in fresh fields. A small proportion become detached to spend a short time in the drift, regaining their foothold in the benthos as soon as they can. With the exception of triclads and trichopterous larvae with a firmly attached case, all species taken in bottom samples were also taken in the drift.

Plate 5 The haunt of the Great silver water beetle *(Hydrophilus piceus)*

(*a*) Water violet (*Hottonia palustris*), one of the water plants on which the beetle feeds.

(*b*) A well-vegetated peaty rhine, typical habitat of the beetle.

Plate 6 Slow-flowing water

(*a*) Part of the Tiverton Canal, Devon, in June. The water flows so slowly that growth of water lilies and other floating water plants is possible.

(*b*) A calm reach of the River Culm, Devon, in June. The river is subject to fluctuations in both depth and speed.

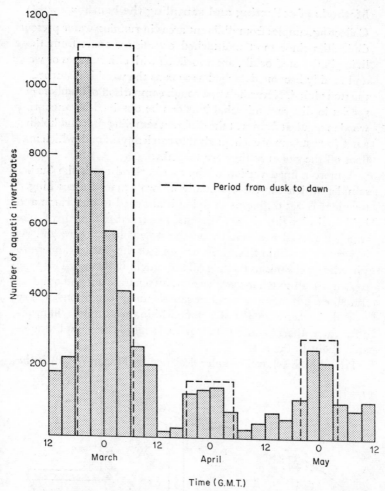

Time (G.M.T.)

Fig. 37. The number of aquatic invertebrates in a Dartmoor stream taken in each sample for a period of 24 hours, in each of the months March to May 1964. (Redrawn from Elliott, 1967.)

D

Methods of collecting and sampling the benthos

Collecting samples from different areas in running water presents difficulties since most unattached organisms, apart from those living in the mud or silt, are associated with either stones or weed and readily become detached as soon as the weed is disturbed or the stone lifted. Nevertheless, a rough comparison of numbers of a species in different microhabitats can be obtained by anchoring wooden quadrat frames at the different sampling sites and holding a net downstream of each quadrat to catch any animals which may float off the stones as they are turned.

A more refined version of the quadrat and net is the Surber sampler which consists essentially of two square frames hinged together. A net of fine mesh nylon is attached to one frame and held vertical so that the net fills out downstream of the other frame which is placed horizontally on the stream bed (figure 38). As larger stones within this frame are turned, the smaller stones are stirred, any organisms floating off are caught in the net. This type of sampler although more efficient than separate net and quadrat frame, cannot make an accurate sample under conditions of very swift flow when removal of a stone within the sampler will probably cause others to shift, so that animals outside the sampling area are caught in the net.

Probably the shovel sampler designed by Macan (1958) is the

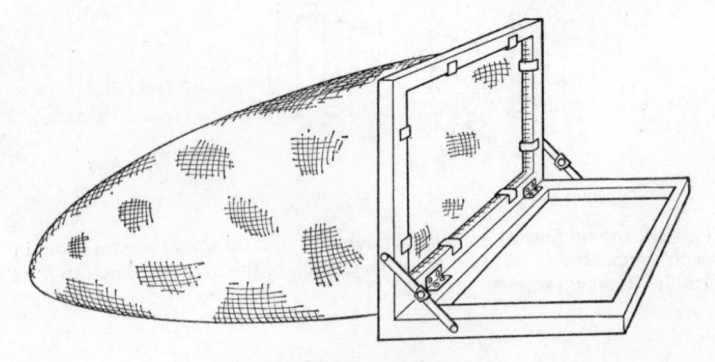

Fig. 38. Surber sampler.

most effective in a stream bed composed of loose stones. This consists of a strong handle at the end of an upright frame to which is attached a coarse net. Outside this, is a larger fine net. The front of the frame has a metal cutting edge which can be driven into the substratum and pushed along until a certain area has been sampled. The stones are retained by the coarse net, while the organisms that are washed off, are caught in the fine net. All the stones are then placed in a strong solution of either calcium chloride or magnesium sulphate and the animals floated off. This method has the obvious, but unavoidable disadvantage of catching the less powerful swimmers which have not had time to flee the approaching shovel.

Measurement of current speed

The bed of any part of a stream is usually uneven and as we have seen any object such as a stone, can alter the rate of flow in its vicinity. Rate of flow also alters with depth and accurate measurement of current speed in the region inhabited by the organisms is difficult since most methods record flow above the stones and not among them.

In open water flow can be measured by means of a small propeller or a Pitot tube, a modification of which is described in Appendix III. Surface rate of flow can be measured in the traditional manner by floating an orange over a measured distance and timing its passage with a stop-watch.

7

Trophic relationships

> '*The large fish eat the small fish;*
> *the small fish eat the water insects;*
> *the water insects eat the plants and mud.*'

Food chains and food webs

This old Chinese proverb describes, without elaboration, the story of what goes on in any stretch of fresh water. In other words a plant is eaten by an animal which, in turn, is eaten by another and this animal may itself be eaten by a third and so on. A sequence such as this is often called a *food chain*. For instance, the microscopic alga *Rivularia minutula* is eaten by the small gastropod, *Hydrobia jenkinsi*, which is the chief food of the larva of the Great silver water beetle (*Hydrophilus piceus*). We can represent this simple food chain, which has but three links, thus:

$$R.\ minutula \rightarrow H.\ jenkinsi \rightarrow H.\ piceus$$

or, in more general terms, as:

$$Plant \rightarrow Herbivore \rightarrow Carnivore$$

But there may be more links than this to a chain. In the above example, the beetle larva may be eaten by a stickleback which itself may be devoured by a larger predatory fish such as a pike. Linear food chains, however, rarely have more than five links.

The general picture which emerges is that even in a small pond the number of microcoscopic plants may run into millions, the numbers of small molluscs such as *H. jenkinsi*, may be very great but there will be fewer beetle larvae and even fewer fish. In other words, the organisms at the beginning of a food chain are relatively abundant while those at the end are few, there being a progressive decrease in numbers but an increase in size of the organisms, between the two ends of a chain. Elton (1966) describes such an arrangement within a food chain as a '*pyramid of numbers*'.

When the food relationships in any freshwater community are investigated, it soon becomes apparent that the feeding arrangements existing between plants and animals are not often simple linear chains such as we have just described. There may be many side links to the chain, for one species may not only use a variety of species as food but may also contribute to the diets of a number of other species. This can more accurately be described as a *food web*, which is really a number of food chains linked together by side chains.

Instead of considering each link in the chain to be composed of one species, it is often best, when studying food webs, to group together the organisms with similar feeding habits.

Figure 39 is a generalized diagram to show the trophic relationships existing in a pond. Dissolved nutrients enter the pond by drainage and seepage from the surrounding land and are incorporated into organic substances by autotrophic bacteria, microscopic plants (phytoplankton), and pondweeds which together are called the *primary producers*. These may die and by bacterial action become incorporated in the ooze at the bottom, or they may be eaten by some consumer. The small drifting animals (zooplankton) such as rotifers, copepods, and some protozoans, feed on phytoplankton and bacteria. The zooplankton, together with some of the larger animals, such as caddis larvae which browse on decaying plant matter and the burrowers, such as the molluscs *Pisidium* spp., some oligochaete worms and midge larvae which feed on the ooze, are called *primary consumers*. The primary consumers are preyed upon by bottom-living animals such as dragonfly nymphs and by plankton predators such as the phantom midge larva (*Chaoborus* sp.). These are termed *secondary consumers*, many of which, like *Chaoborus*, are insects that will leave the aquatic community as adults. The secondary consumers are preyed upon by the larger swimming animals—the *tertiary consumers*—such as fish and water beetles. Finally, primary, secondary, and tertiary

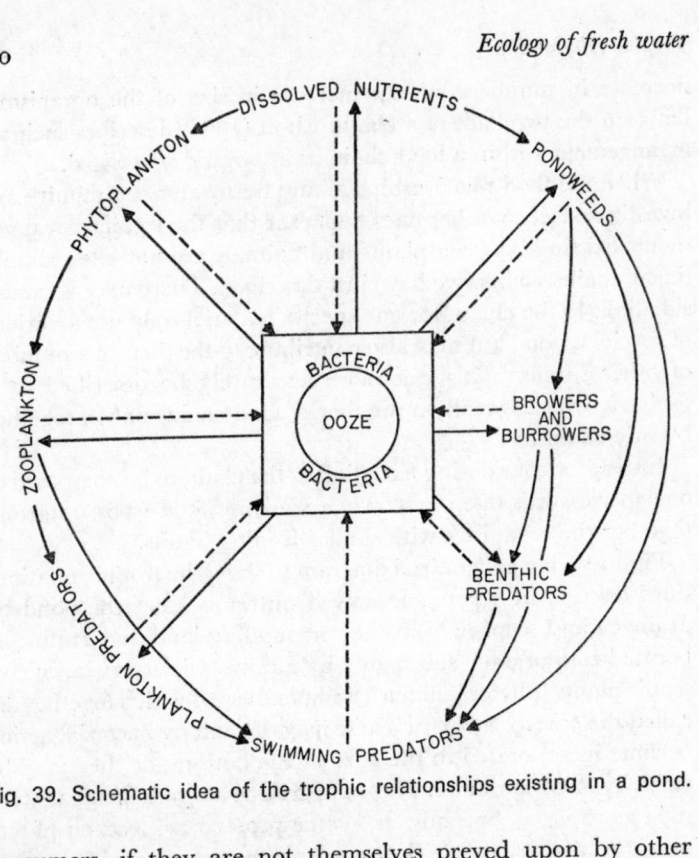

Fig. 39. Schematic idea of the trophic relationships existing in a pond.

consumers, if they are not themselves preyed upon by other animals, will die and contribute to the ooze.

The sequence of events outlined above, can be summarized as follows:

Producers → Primary consumers → Secondary consumers →
                                                    Tertiary consumers

It must be noted, however, that at each level of this chain bacteria break down dead matter. This is, therefore, removed from that level and is not available as food for the next level. There are further complications, for some organisms may not feed only upon food from the level immediately below. Various species of bottom-living water bugs, such as *Plea leachi* and *Ilyocoris cimicoides*, feed on small crustaceans but will also suck up decaying plant and animal debris from the ooze by means of their greatly modified mouth-parts—the rostrum—usually employed to pierce the prey. In this respect they must be considered as primary consumers. Tertiary consumers may obtain food direct from any of the lower consumer groups. A good example is the trout whose diet varies with age. Young trout feed on algae but as they grow larger their diet consists of small crustaceans and later, of larger insects. Many other instances exist of animals which change their diet as they grow older.

One can summarize the situation by saying that the more remote a consumer is with respect to the producers, or primary source of food, the more varied are its food habits likely to be. For this reason all diagrams depicting trophic relationships between one group and another, are somewhat inaccurate especially at the predatory end of a food web.

Predator–prey relationships

Each species has its own seasonal cycle. Besides this, the food of an organism often varies, as we have already remarked, with its stages of growth. For these reasons, the balance between the numbers of predators and their prey is constantly fluctuating. An extreme example of this sort of fluctuation can occur in small bodies of water such as rain-water butts. Pennington (1941) describes the sequence in tubs filled with rain-water in which algae were the first colonists. Rotifers appeared and fed on the algae, becoming so numerous that they reduced the algal population to a level at which they produced insufficient oxygen to support the rotifers. This resulted in the total extinction of the rotifers and the cycle started all over again. This is an example of a very simple chain consisting of two links. In a chain consisting of more links, however, removal of one link by the extinction of a species, may alter the balance of the chain, not only at the level where that species is exterminated but at all levels.

The density of a predatory species tends to be greatest after the point of maximum density reached by its prey (figure 40). More usually as the prey population decreases, the number of predators also decrease to a level where the prey recover and begin to multiply. With the renewed food supply the predator species begins once more to increase. These fluctuations can continue out of phase with one another through several cycles.

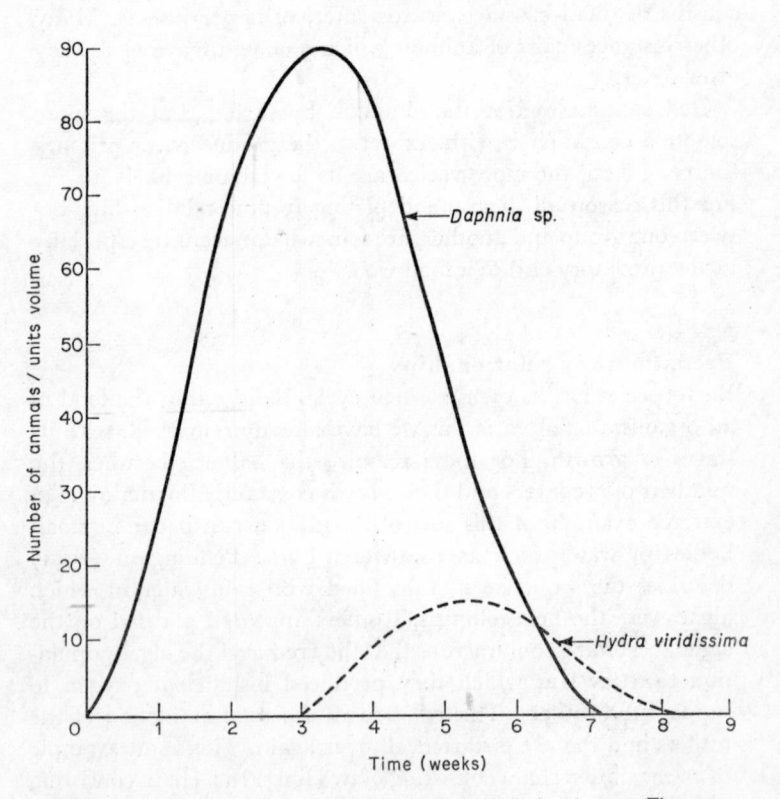

Fig. 40. The density relationship between predator and prey. The prey population (*Daphnia* sp.) increases until the predator (*H. viridissima*) begins to multiply, eventually exterminating the prey and itself declining in numbers to extinction.

Methods of tracing food chains

Within the confines of an aquarium, direct observation of who eats what may be possible. For instance, one may be able to see that a caddis larva is feeding on pondweed and that a stickleback eats the caddis larva. But this will only give a very approximate answer.

A more exact estimation can be obtained by the analysis of gut contents. This involves the collection of a large number of individuals of each species and the careful examination in each case, of separate parts of the gut. Although it may be possible to identify the hard parts of organisms present in the gut, such as the skeletal remains, chitin, and the like; parts of soft-bodied animals may escape notice, for it is impossible, by such means to identify body fluids and plant juices which are rapidly assimilated after ingestion. In any case the results obtained by such methods do not ustify the slaughter of the large numbers of animals required for this method of analysis.

Recently the *precipitin test* has been used with success in tracing food relationships in certain species of aquatic organisms. Reynoldson *et al.* (Reynoldson and Young, 1963; Young, Morris, and Reynoldson, 1964) have used this method in tracing the diet of the flatworm, *Dendrocoelum lacteum*, with some success. Gut analysis of this species revealed undigested setae identified as belonging to oligochaete worms. Direct observation of the feeding habits of *D. lacteum* showed, however, that it preferred the water louse, *Asellus*, ingesting only the soft tissues and body fluids. The precipitin test was applied by inoculating caged rabbits with cell-free extracts of *Asellus* to produce an antiserum. Smears of *D. lacteum* were then made by crushing the animals on filter-paper, drying it rapidly over phosphorus pentoxide and extracting the smears in normal saline which were then centrifuged. A small amount (0·02 ml) of the extract was drawn into each of several capillary tubes followed by an equal volume of antiserum. After keeping the tubes at room temperature for two hours, the formation of a precipitate of antigen and antibody at the confluence of the two liquids, indicated that the food of *D. lacteum* consisted largely of

Asellus. This result would not have been revealed by gut analysis alone.

More recently radioactive isotopes have been used in tracing terrestrial food chains. Plant foliage of a single species was 'labelled' by spraying with phosphorous-32. Animals living in the vicinity of the plants were then tested for the presence of ^{32}P. Any species showing radioactivity must have been either directly or indirectly dependent on the plants as the original source of food. Any radioactive substance which can be used in this way to 'label' the primary food source and which can be subsequently traced through other organisms, can be of value in elucidating the stages of a food chain. Such methods might also be possible in tracing an aquatic food chain.

Pyramids of biomass

In working out food relationships it may be possible to discover the number of producers which support a given number of primary consumers which, in turn, support a given number of secondary consumers, and so on. But by using this kind of approach for the comparison of two ecosystems, we might soon find ourselves in trouble if we tried to equate say, the number of caddis larvae feeding on pondweed in an aquarium with the number of cows feeding on swedes in a field. If, however, their weight or *biomass*, is used instead of the number of the organisms composing the different trophic levels, it is possible to construct a *pyramid of biomass*.

Philipson (1968) constructed such a pyramid for Malham Tarn (figure 41). This was done by taking samples of plant and animal material from different regions of the tarn. The material was then sorted into categories according to which trophic level it belonged—primary producers, herbivores, or carnivores. The mean biomass of the samples in each category was then calculated and expressed as grammes wet-weight per square metre of tarn surface.

The pyramid of biomass shown in figure 41, is what might be expected, for a higher weight of producers is required to give rise

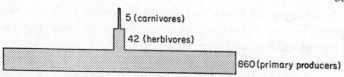

Fig. 41. Mean biomass pyramid (expressed as g wet weight/m^2) for Malham Tarn. (After Philipson, 1968.)

to a lower weight of consumers. Yet this is not always the case especially in the relationships existing between the phyto- and zooplankton of a pond. Calculations of biomass made during spring and early summer may reveal an inverted pyramid, the plant plankton giving rise to a larger weight of animal plankton. This can happen at certain seasons of the year in the sea, when the ratio of dry weight of zooplankton to phytoplankton may be as much as 5:1.

Aquatic plants grow at different rates and the time taken to complete their life-cycles varies. Bacteria and algae may divide several times per day, while pondweeds develop one generation a year. When it is realized that many crops of phytoplankton are synthesized from the original supply of nutrients, that they are consumed, decomposed, and resynthesized several times during the year, it is hardly fair to compare the total annual production of phytoplankton with that of pondweeds.

Pyramids of biomass, therefore, can only show the amount of material present at one particular time, or what is called the *standing crop*. Such pyramids cannot tell us what the total production of material is nor the rate at which it is produced. For these reasons it is necessary to look for another means of assessing food relationships.

Energy balance

All living things require energy for their life processes. The ultimate source of energy is that which is radiated by the sun. During the process of photosynthesis, green plants are able to combine carbon dioxide, water and light energy, storing the energy as carbohydrate.

In order to follow the part played by energy within an eco-system, we must know more about the energy used and lost by an organism during its life processes.

Animals feeding either directly (herbivores) or indirectly (carnivores) on plant tissues, oxidize the carbohydrate and liber-ate carbon dioxide, water, and energy.

The act of feeding can more accurately be described as *ingestion* to distinguish this process from assimilation. This is the absorption of that portion of food ingested which is actually taken into the body tissues. The remainder is got rid of as faeces and of that which is assimilated, part is stored while part is metabolized to liberate energy.

Energy manifests itself as heat and heat is measured in *calories* (1 calorie = 4·2 joule), one calorie being the amount of heat required to raise 1 g of water 1 °C. The number of calories a substance contains can be calculated by completely combusting a known dry weight of the substance in an insulated metal container, or bomb calorimeter, and measuring the heat produced. This is called the calorific value of the substance.

The complete combustion of 1 g of carbohydrate liberates 673 000 calories of heat and the process of such an oxidation can be represented by the equation:

$$C_6H_{12}O_6 + 6O_2 = 6CO_2 + 6H_2O - 673\ 000\ \text{calories}$$

The same amount of energy is liberated as the result of the com-bustion of 1 g of any kind of carbohydrate. Fats liberate rather more energy and proteins produce less.

Investigation of the energy balance of an animal will, therefore, involve measurements of the calorific value of its ingested food as well as that of the assimilated food, faeces, stored food, metabolized food, and the amount of energy liberated as heat during the metabolic processes. The ratio of stored food to assimilated food is a measure of the efficiency of the organism. In the water flea (*Daphnia*), for instance, the ratio of food stored to that which is assimilated can be as much as 50 per cent, a higher figure than for most organisms.

8

The flow of energy through an ecosystem

Of the radiant energy entering the plants which form the lowest trophic level of an ecosystem, about 50 per cent is lost by radiation. The remainder is transformed into carbohydrates, fats, and proteins. Of this energy, about 30 per cent is lost during the respiratory processes of the plants themselves. Despite this, a large amount of energy is stored as plant tissue to be made available to the organisms at other trophic levels. The herbivores feeding on the plants partly break down the organic matter they ingest, storing some of the energy intake in their bodies and giving off some in the form of heat during their metabolic processes. The same thing happens at higher levels of the system and the *productivity* of an organism or of a group of organisms, composing one trophic level, can be expressed in terms of energy units. The important point is that this 'lost' energy, unlike chemical substances, does not recirculate within the ecosystem. It therefore offers a valid means of comparing the total productivity either of a single organism, a group of organisms composing one trophic level or of a whole ecosystem.

Odum (1957) working at Silver Springs, Florida, measured the total energy entering the spring as well as the amount of energy passed on by one trophic level to another and the amount of heat lost by each level. From these measurements it was possible to construct an energy flow model (figure 42) for the whole system.

It is clear from the diagram that the energy entering the system is almost entirely the result of the photosynthetic activities of the plants, while the total energy lost to the system is made up of

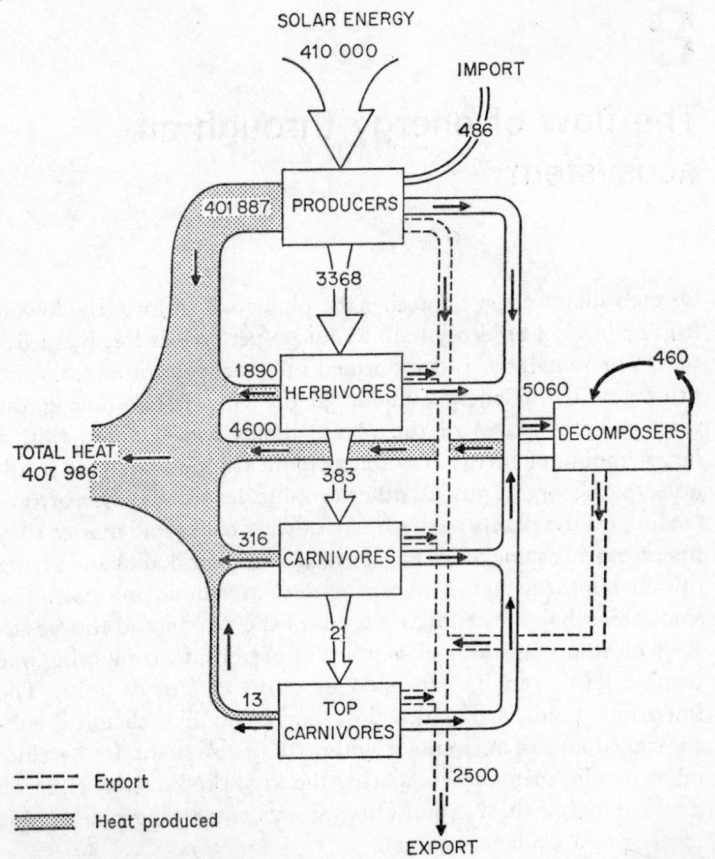

Fig. 42. Schematic drawing of the flow of energy in Silver Springs, Florida. Figures in kcal/m² year. (Adapted from Odum, 1957.)

the sum of all the energy losses due to the metabolic activity of the plants and animals composing the community.

Decomposers

In the past the part played by the detritus feeders and decomposers has been much neglected, but in many types of ecosystem they have an important function to perform. Ultimately, most of the food consumed at the different trophic levels, is dealt with by these organisms which, through their metabolic processes, break down waste and decaying matter liberating nutrient inorganic matter which once more becomes available to the plants.

Productivity

Recent studies of different ecosystems by Odum (1957), Lindemann (1942), and others, have shown that the efficiency with which energy is transferred from one trophic level to another varies with different systems. Natural ecosystems are complex and the investigation of a complete ecosystem from the trophic-dynamic point of view can be an enormous task.

Teal (1957) chose to study a small spring, called Root Spring in Massachusetts, which was about 2 m in diameter and 10 to 20 cm deep. Being small and containing a relatively simple community of plants and animals, the spring was virtually a 'laboratory ecosystem'. Another important factor was that the bottom of the spring had a thick deposit of mud, covered with a layer of dead leaves in which most of the organisms dwelt. The most abundant animals were detritus feeders such as *Limnodrilus* sp., *Asellus* sp., *Pisidium* sp., *Physa* sp., *Crangonyx* sp., and *Calopsectra* sp.

Since most of the plant material forming the food of the herbivores entered the spring as debris (dead leaves and other parts of overhanging terrestrial vegetation), an estimate of the energy entering the ecosystem involved the measurement of energy produced by the breakdown of the debris as well as measurement of the gross production by the plants living in the spring, such as diatoms, filamentous and colonial algae, and the duckweed, *Lemna minor*.

To catch falling leaves and other plant debris, traps were put beside the spring and the calorific value of the debris entering the spring in this way was calculated to be 2350 kcal/m² year. The gross production of energy by the plants, all of which were either microscopic or very small, was measured by the dark and light bottle method (Odum, 1963) and expressed as kcal/m² year. The following results were obtained:

	kcal/m² year
Gross productivity of plants	710
Estimated respiration of plants	55
Net productivity of plants	655
Energy produced by plant debris	2350
Total energy entering the ecosystem	3005
Energy not used by herbivores but deposited	705
Energy utilized by herbivores	2300

Energy transfer

As we have already seen, energy is continuously entering and leaving at each trophic level of an ecosystem. In the case of Teal's spring, the figure of 2300 kcal/m² year represents the amount of energy actually consumed by the herbivores. But of this only 655 kcal was produced within the ecosystem, the rest entered from *outside* as debris—an unusual state of affairs.

Teal went on to discover what use was made by the herbivores of the energy they consumed and how much they passed on to the carnivores. This involved the estimation of numbers for each species and then from their weights, determining their calorific values. The amount of heat lost by each species through metabolism was discovered by measuring their oxygen consumption and converting this figure to mg of oxygen per g of body weight per unit time. In this way energy balance sheets were constructed for each species and from this information it was possible to gain an idea of the way in which energy flowed through the ecosystem.

The important fact emerging from Teal's work is that most of the animals in Root Spring were decomposers feeding on debris which entered the spring from outside. Odum (1957) also found an abundance of decomposers at Silver Springs but in this case their food was produced *within* the ecosystem.

There appear to be two kinds of food chain within each of these systems: a grazing chain and a detritus chain which can be represented thus:

Living plant material → Grazing herbivores → Carnivores
Dead plant material → Detritus herbivores → Carnivores

Plate 7 Exton stream, Devon, in June

(*a*) The weir and pool. In the foreground some of the large blocks of concrete washed into the stream during the floods of 1960.

(*b*) A deep and shallow reach.

Plate 8 Stream and river currents

(*a*) A swift reach of the River Otter, Devon. Note the large stones in the river-bed which create differences of current in their vicinity.

(*b*) Students using a current-meter to record rate of flow where populations within a quadrat frame were being examined in Exton stream, Devon.

These two food chains are not always isolated from one another. Odum's work indicates that the dead bodies and faeces of animals which formed part of the grazing chain, became incorporated in the detritus chain.

The importance of the decomposers in an ecosystem is also shown by Odum's results where 5060 kcal/m² year flowed through the decomposers while only 3368 kcal/m² year went through the grazers (figure 42).

Energy losses in aquatic systems

Besides the energy losses due to metabolism, there are others in an aquatic ecosystem which must be taken into account. Many organisms with aquatic larvae are terrestrial as adults, thus when drawing up an energy balance sheet for Root Springs, Teal had to take this into consideration. For instance, in the case of the midge larva, *Calopsectra dives*, monthly counts were made of the number per m² of larvae and emerging adults. These revealed that the larval standing crop grew from zero in April to 57 000, or 87·6 kcal/m² in August, but fell to zero once more by October. Since the energy change in the standing crop of *C. dives* over the year was nil, none of the energy consumed was used to increase the overall standing crop of the species. But in considering the total energy contributed by *C. dives* to the ecosystem, besides an estimate of the monthly respiratory heat losses of the standing crop, account had to be taken of other factors. On the debit side, a calculation had to be made of the number of adults leaving the spring on emergence, converting their biomass into kcal/m², while on the credit side, was the calorific value of the larvae dying each month and that of their cast skins deposited in the spring.

Ecological efficiency

The Law of Conservation of Energy states that:

> *Energy may be transferred from one form into another but is neither created nor destroyed.*

The flow of energy through an ecosystem illustrates this law inasmuch as radiant energy from the sun enters the system to

E

become converted into energy stored in living tissues or used in
mechanical work. Most of the radiant energy available to a
system is never incorporated, while much of the incorporated
energy is dissipated as heat.

The question now remains of how efficiently energy is transferred from one trophic level to another. In an ecosystem the
amount of energy made available to a predator at the next trophic
level divided by the amount of food ingested by the prey is a
measure of the *ecological efficiency* of those levels of the ecosystem.
This can be represented in this way:

$$\frac{\text{Energy contained in yield to predator}}{\text{Energy content of food ingested by prey}} = \text{Ecological efficiency.}$$

Philipson (1968) in his survey of Malham Tarn, found that the
ratio of primary consumers to primary producers was 1 : 20 and
that of the secondary consumers to the primary consumers was
1 : 8 (figure 37), both being derived only from measurements of
standing crop biomass. Since herbivores have a less efficient
digestive system than that of carnivores, one would expect the
former of the two ratios to reflect this greater inefficiency.

Few measurements have so far been made to determine the
ecological efficiencies of natural systems and figures range from
5 to 30 per cent, but Slobodkin (1960) considers 10 per cent to
be fairly consistent among a range of conditions investigated.

Theoretically, in a perfect ecosystem, there should be sufficient
organisms at each trophic level to consume during the year the
organisms of the level below, down to a point where they are
never numerous enough to run short of food nor depleted in
numbers to a point where any of their food goes to waste.

9

The influence of man on freshwater communities

Various factors have already been described which limit populations within a community as well as their ability to spread to others. To this list we must now add man who is one of the most potent factors in altering the structure of a community. There are so many ways in which the balance can be upset by the workings of man, supposedly for his own benefit, that it may be as well to discuss how some of these can influence freshwater communities.

Pollution
This is a term that can be rather widely applied to include substances introduced into an environment which are potentially harmful to, or which interfere with man's own use of his environment.

Pollution of water in various ways by waste products is of great importance, but we can make a fundamental distinction between two types of waste product which are causes of pollution: (1) those which involve either an increase in volume or rate of introduction of substances which are already present in the water, and (2) those products, such as poisons and chemicals, including atomic waste, which are not normally present in a natural ecosystem.

In the first instance, we are dealing chiefly with organic waste, such as sewage and the ordinary minerals which are present in low concentrations, in all ecosystems. These pose a lesser problem, since biological means of controlling their amounts can be applied. However, it must always be remembered that what man is trying to do for himself, is to bring within his own limits of tolerance, the factors causing pollution, and these may not be the tolerance limits of other organisms. Fish, for example, require quite large amounts of oxygen in solution, while the oxygen demands of crustaceans such as *Asellus*, may be less and that of certain bacteria even less still.

Sewage

A century or so ago, most rivers flowing through towns and villages, served not only as the water supply of the inhabitants, but also as the town sewer. What happened downstream was of no concern to those tipping sewage into the river higher up. The growth of industrial areas, and with them the enormous rise in populations, might well have meant their annihilation, had not two great advances come about at roughly the same time. The work of Louis Pasteur and others, showed that diseases such as cholera and typhoid were waterborne and secondly the Public Health Act of 1875 and the Rivers Pollution Prevention Act of 1876, which although they were not immediately successful, paved the way for effective control of water supplies.

Sewage, when entering a stream or river in an untreated state, reduces the amount of oxygen in the water due chiefly to the activities of bacteria in breaking down the raw sewage into simple compounds such as nitrates, sulphates, and phosphates. Since all of these are compounds involving oxygen, sewage effluents tipped into a river will soon use up the available oxygen dissolved in the water. When all available oxygen has been used, decomposition can still proceed but is brought about by anaerobic bacteria, poisonous compounds such as ammonia and hydrogen sulphide, being the end products. It is a salutary thought that the sewage produced by a single human being daily gives rise to an oxygen demand of about 115 g or the equivalent of oxygen contained in 10 000 l of oxygen-saturated water.

The speed with which bacterial breakdown of sewage takes place is increased with temperature. This is important when the amount of pollution per unit volume of water is considered. If the water is warm near a sewage outflow, decomposition is rapid and will take place within a short stretch of water, thereby concentrating the area in which there will be a depletion of oxygen. The oxygen threshold may fall sufficiently low to cause the death of fish and other organisms in the area. Colder water, on the other hand, will mean that decomposition will extend over a greater length of the river. Consequently, the sewage will become mixed with a larger volume of water and the danger of oxygen shortage will be decreased. Obviously the volume of water and the rate of flow have to be taken into consideration in determining the danger point. To land-living, water-drinking human beings, such a danger point can be reached if the level of water-borne disease-causing bacteria rises to a certain level.

There are various methods for the treatment of raw sewage and a vast literature exists on the subject. There is only space here to discuss one method, and probably that which is the most popular in Britain. This employs the sprinkling and trickling filter in which the organic matter is first broken down by bacterial action and the liquor is then distributed by rotating arms through beds of broken stone, coke, or clinker. These offer a very large surface which is colonized by a film of slime consisting of algae, bacteria, and protozoa. The first colonists are the bacteria and on the surface, filamentous algae develop, mainly the blue-green species. The film also supports many protozoa including fixed species such as *Vorticella* and *Epistylis* (figure 43(a)), as well as free-moving forms. As the sewage continues to feed these organisms, the film increases and a mat of fungi and algae can develop into a thick leathery layer which would seal the surface. In a properly functioning filter, such a growth is checked by the feeding activities of worms and insect larvae which graze on the film surface and beneath it. Much of the film is then incorporated in the bodies of these organisms and either leaves the filter in their excreta or when the insects emerge. The shedding of the film in this way, is a continual process.

The special nature of the filter-bed makes it habitable by only a few invertebrates. Two species of oligochaete worms, *Lumbricus rubellus* and *Lumbricillus lineatus* (figure 43(b)) are present in large numbers. The latter is a particularly successful colonist because, unlike other enchytaeids, it can attach its eggs to the film. Other species are three moth flies (*Psychoda*) and the Window fly (*Anisopus fenestralis*) whose larvae also colonize the filters. The numbers of species, however, are limited by the strength and nature of the sewage. Nevertheless, the efficiency of the filter-beds

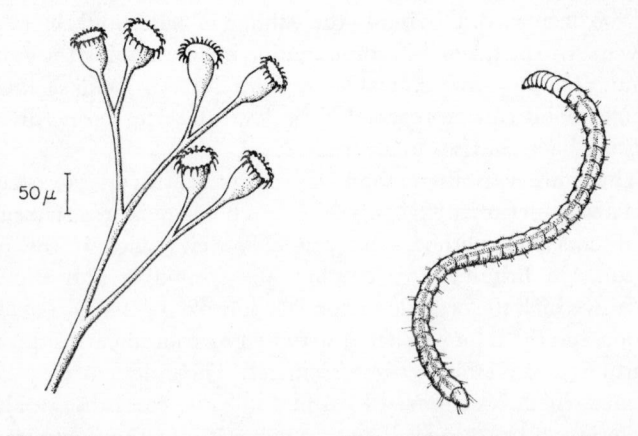

50 μ

(a) *Epistylis* sp.   (b) *Lumbricillus lineatus* 25mm

Fig. 43. Two inhabitants of sewage. (a) *Epistylis.* A fixed, colonial ciliate which is non-retractile. It lives in the film surrounding the sewage bed and is one of the organisms causing the malfunctioning of the bed by the formation of a thick, sealing film. (b) *Lumbricillus lineatus* is an oligochaete worm which grazes on the sewage film helping to check its growth.

depends upon the presence of these grazing organisms which form the third link in the sewage food chain.

The final effluent passes through sedimentation tanks which remove the fine organic mud, after which it should be fit to drink. The river Thames, once nothing better than a glorified sewer, is now said to be drunk, excreted, and drunk once again, several times over between source and estuary, but the addition of large quantities of chlorine, makes London drinking water somewhat unpleasant.

Other types of pollution

Besides sewage a number of waste products are discharged into our rivers by factories and trades of different kinds, such as woollen mills, breweries, and slaughter houses, often with little ill effect since they are of organic origin and can therefore be treated in the same way as sewage.

Others, however, can produce poisonous substances. Glue factories and tanneries as well as heavy industries which produce waste metals such as copper, lead, iron, and zinc, all have a direct affect on gill-breathing organisms.

(i) *DDT* One of the most serious substances entering water is DDT (dichlorodiphenyltrichloroethane) since it can produce cumulatitive effects. It is often used in sheep dips or as a control spray against certain insects and can deplete the insect fauna of a stream and, indirectly, fish feeding on the insects and birds feeding on the fish. + Diagram from journal.

(ii) *Lead* Carpenter (1926) surveyed a stream in Cardiganshire draining a valley in which there were a number of disused lead mines. The survey was made in 1921 and 1922, four years after some of the mines had been reopened for a short period during the First World War. Only fourteen species of animals were found, most of them insects, but there was a notable absence of trichopterans. Analysis of the water showed 0·2 to 0·5 ppm of lead. The following year there was a gradual recovery when the concentration of lead had fallen to 0·1 ppm and twenty-nine species, including trichopterans and the flatworm, *Polycelis nigra*, were found. Surveys made by others in 1931 and 1932 in the same stream, discovered 103 species including fish and molluscs, when the concentration of lead had fallen to 0·02 ppm rising to 0·1 ppm after heavy rain.

These detailed surveys made on one stream polluted by lead, indicate four definite stages in the recovery of the stream fauna from (*a*) no life at all to (*b*) a small number of insect species, excluding trichopterans to (*c*) a richer fauna including oligochaete worms, turbellarians, molluscs, and fish. Throughout these faunal stages there was a succession of plants starting with a few algae, mosses, and liverworts and ending with the appearance of higher plants.

(iii) *Nitrates and phosphates*   These are substances found naturally in all lakes and rivers but amounts of these salts can be increased enormously by run-off from land treated with chemical fertilizers. In enclosed bodies of water such as lakes and reservoirs, high concentrations can cause a sudden increase in the quantity of algae which will turn the water into a turbid pea soup rapidly depleting its oxygen content. This is known as *entrophication* and has occurred periodically in one or two reservoirs in England and Wales. This is a form of water pollution difficult to control but needing a careful watch.

(iv) *Detergents*   As washing agents, detergents differ from ordinary household soap in being manufactured from various synthetic chemicals which, unlike soap, are not precipitated but are soluble in water. Sewage from houses and factories using detergents which discharge into rivers can cause persistent foam to accumulate wherever there is slight turbulence of the river and usually this occurs some distance below where the sewage works discharges its effluent. This is because foaming will not take place in dirty water and it is not until some self-purification of the water has occurred that production of foam is possible. Indeed, pollution of our rivers might have gone unnoticed had it not been for the vast quantities of foam produced in some of them. As little as 0·1 ppm of detergent, reduces the rate at which a river takes up oxygen by nearly 50 per cent. Trout and other fish can be affected by concentrations of detergent of as little as 1 ppm. This is the effect on rivers. It only requires slight imagination to realize the effect of the discharge of detergents from cottages and farmhouses upon the fauna of an upland stream. Even a small amount of detergent can put a domestic septic tank out of action.

Because detergents are resistant to bacterial action they also reduce the efficiency of sewage works. Hence, should they have to cope with a high concentration they have to be extensive.

Assessing pollution

Of prime importance to man is a pure supply of water. This involves the control of harmful organisms, normally present in water, below the level at which they can do damage within the alimentary canal or other parts of the body.

Chemists are able to make a number of quite simple tests for the presence of harmful chemicals and poisons, and bacteriologists can make certain standard tests for the presence of pathogens. But the fact remains that there may easily be intermittent pollution, which escapes notice if the samples happen to be taken at the wrong time or place. Nevertheless, certain standard tests are regularly made which assess the degree of pollution.

Chemists can measure the amount of oxygen which disappears in a given time at a certain temperature and they assume this to be proportional to the amount of organic matter present. This is called the biochemical oxygen demand, or B.O.D. But this is not the whole answer, for human pathogens affecting the alimentary canal, present only in small concentrations, can still be harmful without significantly altering the B.O.D. Supplies of drinking water have, therefore, to undergo bacteriological tests to detect the presence of organisms known to occur in the human intestine. This could involve lengthy analysis, since there are a number of water-borne bacteria falling within this category. One of these is *Escherichia coli* and is used in official tests as the indicator, because so far as we know, it does not occur naturally in water. Its presence, therefore, is certain proof of recent organic pollution. The official standard laid down is that water containing more than 1–5 coliforms per 100 ml is regarded as unfit for drinking.

Summarizing the very complex question of water pollution and the diverse tests for pollution which can be applied, we can say that chemicals and poisons are detected by the chemist who can also determine the amount of pollution due to the presence of organic matter. The biologist is concerned with level of concentration of oxygen and its effect on the fauna. More recently the importance of the biologist within the field of water analysis, has come to be realized, with the result that an increasing number of water boards are including biologists on their staffs. There is no more important field of applied science than that which can now be called 'environmental engineering' in which the ecologist and the engineer combine in an effort to cope with the ever-increasing volume of man-made wastes.

Increasing productivity

Man is much concerned with devising ways of augmenting his food supply by increasing the numbers and weight of edible freshwater fish.

Under natural conditions, the growth of fish depends not only on the amount of available food but also upon temperature. Different species of fish grow at different rates. The average weight of a yearling carp is about 55 g and it will increase in weight to about 340 g in its second growing season, while a yearling trout will increase in weight during this period by about half this amount.

Much experimental work has been done on the growth of brown trout and the conclusions to be drawn are that water temperature is the main environmental factor influencing growth rate, maximum growth being achieved at 12°C. Above this temperature the rate decreases. Fry (1957) suggests that this may be due to the inability, at higher temperatures, of the respiratory capacity of the fish to meet its respiratory requirements.

The type of food chain existing in a pond or lake is of great interest in deciding how best to increase a stock of fish, the nature of the bottom, and the amount of light being important points. Where there is a good mixture of organic matter and of inorganic material such as silt, in a state of fine division, decomposition will be more rapid. This type of bottom will retain any minerals added to increase the growth of algae, releasing them slowly. A peaty bottom usually lacks calcium which, in turn, means a very slow rate of decay. Furthermore, any added minerals may be washed out of the pond before the plants have been able to make use of them. The addition of lime to such a substratum, will increase the rate of decomposition.

The addition of phosphate at the rate of about 200 kg per hectare, either in the form of basic slag, bone meal or superphosphate, has proved beneficial provided that the calcium requirements of the water have been taken into consideration. The amount of calcium which appears to be necessary for the satisfactory functioning of a food chain, is approximately 20 ppm. Besides phosphate, the addition of potassium in the proportion of about 1 part of potassium to 2 of phosphate, has been used on many European fish farms, besides frequent small applications of organic manures.

These methods of increasing fish stock have been concerned with the beginning of the food chain by increasing plant growth for the consumption of the herbivores, and with the end of the chain. But a knowledge of the intermediate links and of the side-chains must obviously be considered in the cultivation of each species of fish in different ponds or lakes.

Conservation of our rivers and streams

Food chains exist in large and small freshwater ecosystems and, as we have seen, the destruction of one link in the chain can cause a complete alteration of the relationships within an ecosystem. In order to preserve the stability of an ecosystem it is of the greatest importance that the physical and chemical structure of our natural waters is maintained and that the biotic influences which man can so easily exert, are kept in check.

As a nation we are becoming increasingly aware that time is running out and that urgent steps must be taken to preserve our lakes and rivers from the more obvious causes of pollution as well as the indirect causes. It is encouraging to know, for instance that, by government order, DDT as an insecticide is to be phased out and its place taken by other chemicals such as 'Sevin' which has a much lower toxicity and which, unlike DDT, is rapidly metabolized and is not stored in animal tissues.

Individual responsibility in preventing the pollution of smaller bodies of water on private land by chemicals and harmful effluents, is an important contribution to the conservation of the flora and fauna of our natural waters, which it is possible for everyone to make.

buy green products.

Appendix I

Table of the solubility of oxygen in chloride-free water at various temperatures when exposed to water-saturated air at a total pressure of 760 mmHg and partial pressure of oxygen at 160 mmHg. (Dry air is assumed to contain 20·9 per cent oxygen.)

°C	ppm	°C	ppm	°C	ppm
0	14·62	17	9·74	34	7·2
1	14·23	18	9·54	35	7·1
2	13·84	19	9·35	36	7·0
3	13·48	20	9·17	37	6·9
4	13·13	21	8·99	38	6·8
5	12·80	22	8·83	39	6·7
6	12·48	23	8·68	40	6·6
7	12·17	24	8·53	41	6·5
8	11·87	25	8·38	42	6·4
9	11·59	26	8·22	43	6·3
10	11·33	27	8·07	44	6·2
11	11·00	28	7·92	45	6·1
12	10·83	29	7·77	46	6·0
13	10·60	30	7·7	47	5·9
14	10·37	31	7·5	48	5·8
15	10·15	32	7·4	49	5·7
16	9·95	33	7·3	50	5·6

Appendix II
Methods of measuring dissolved oxygen

(a) Winkler method
The determination of oxygen by this method involves four stages:

(1) When manganous hydroxide is added to a known volume of water containing oxygen, a brown precipitate of manganic hydroxide is formed as follows:

$$2Mn(OH)_2 + O + H_2O \rightarrow 2Mn(OH)_3$$

(2) This precipitate is dissolved in a non-oxidizing acid (chlorine-free hydrochloric acid or ortho-phosphoric acid) with the formation of manganous chloride as follows:

$$2Mn(OH)_3 + 6HCl \rightarrow 2MnCl_3 + 6H_2O$$

(3) By the simultaneous addition of potassium iodide, an equivalent quantity of iodine is liberated as follows which turns the liquid a clear brown colour:

$$2MnCl_3 + 2KI \rightarrow 2MnCl_2 + 2KCl + I_2$$

(4) The iodine is then titrated against a standard solution of sodium thiosulphate, using starch solution as an indicator, the reaction being represented thus:

$$2Na_2S_2O_3 + I_2 \rightarrow Na_2S_4O_6 + 2NaI$$

Reagents required
A. 40 per cent manganous chloride solution.
B. 33 g sodium hydroxide and 10 g potassium iodide dissolved in 100 cm³ distilled water.

C. N/80 sodium thiosulphate solution.
D. Concentrated hydrochloric or orth-phosphoric acid.
E. Starch solution. Boil some starch in distilled water. Cool some
   of the solution before use.

Procedure

(1) A stoppered bottle completely filled with the sample of water
   to be tested, should be used so that the sample does not come
   into contact with air. A convenient size of bottle is one of
   70 cm³ capacity.

(2) Remove stopper and, for a 70 cm³ sample, add 5 cm³ of A
   and 0·1 cm³ of B. (For volumes greater or less than 70 cm³
   calculation of the amounts of A and B to be added must be
   made.) Ensuring that no air bubbles are included, replace
   stopper and shake. Leave to stand for 5 minutes when a
   brown precipitate of manganic hydroxide will appear and
   oxygen in the sample is now fixed. This operation is best
   carried out in the field.

(3) Add about 2 cm³ of B, replace stopper and shake. Iodine will
   be liberated.

(4) Titrate 25 cm³ of the liquid against solution C to which a
   drop or two of freshly prepared indicator E has been added.
   Repeat procedure using a second sample.

Calculation of oxygen content of sample

$$1 \text{ cm}^3 \text{ N/80 Na}_2\text{S}_2\text{O}_3 \equiv 0\cdot1 \text{ mg O}_2$$

The amount of oxygen/l can be calculated as follows:

$$\text{mg oxygen/l} = \frac{V \times 0\cdot1 \times 1000}{v}$$

where V = volume of sodium thiosulphate used and v = volume
of sample.

$$1 \text{ cm}^3 \text{ N/80 Na}_2\text{S}_2\text{O}_3 \equiv 0\cdot0001 \text{ g oxygen}$$

therefore $1 \text{ cm}^3 \text{ N/80 Na}_2\text{S}_2\text{O}_3 \equiv 0\cdot0001 \times 22\,400 \text{ cm}^3$ oxygen therefore $\dfrac{V \times 0 \cdot 0001 \times 22\,400 \times 1000}{32 \times v} = \dfrac{V \times 70}{v}$

$=$ cm³ oxygen/l at s.t.p.

Note (1) N/80 sodium thiosulphate must be standardized against standard potassium dichromate or potassium permanganate. The shelf life of this solution is short and a formation of precipitate indicates that it is no longer fit for use.

(2) Starch solution must be freshly prepared for each series of titrations.

The above method is that described by Dowdeswell (1959)

(b) The Protech portable dissolved oxygen and temperaturemeter

For more accurate work or where a continuous record is required, the Protech portable oxygen meter, Model SM-121A can be used.

This type of instrument records direct readings of dissolved oxygen continuously in any type of flowing water. It has a dissolved oxygen range of 0–15 ppm with an accuracy of + or −0·1 ppm and a temperature range of 0–40°C with an accuracy of + or −0·5°C. The instrument can also be used to make continuous readings of oxygen consumption by small organisms (Mann, 1965).

In addition, this particular model can be used in conjunction with a carbon dioxide electrode to record dissolved carbon dioxide, since both electrodes record over the same period of time.

Further particulars can be obtained from Protech Advisory Services Ltd., Gable House, 40 High Street, Rickmansworth, Herts. The paper by Nicholls, Shepherd and Garland (1967) should also be consulted.

Appendix III

Measurement of flow by means of a current-meter

Various types of flow-meter exist, the more expensive ones involving electrical recording devices.

Where measurements of flow are to be made at different points for comparative purposes and continuous readings over a period of time are not required, a simple apparatus can be constructed using two Pitot tubes.

The U-tube is half filled with a coloured fluid and joined, on either side, by means of rubber tubing, to a Pitot tube. A right-angle piece of copper tubing is attached to the end of each by means of rubber tubing and the whole apparatus is fixed to a board. Behind the U-tube is fixed a piece of graph paper and a scale behind the tubes, to indicate the depth of the tubes beneath the surface.

To take a reading of current speed at a certain depth, the meter is held so that one of the copper tubes points upstream and the other down. This will cause a difference in air pressure in each of the long tubes, which results in the displacement of the liquid in each arm of the U-tube. The difference in height of the liquid in each arm of the U-tube, measured on the graph paper, is proportional to the speed of the current at that depth.

For calibration the apparatus should be held *just* beneath the surface of the water, in a stretch of river where the current is uniform. By taking readings of the differences in the height of the liquid in the two arms of the U-tube at different sites where there are different current speeds (judged by floating an orange and timing the distance travelled over a known length), a graph can be plotted of height difference and current speed in m/sec which can be fixed to the back of the meter board.

In order to contain the coloured liquid in the U-tube when the board is carried horizontal, it is best to attach spring clips to the rubber tubing.

This method of constructing a flow-meter is described by Dowdeswell (1959).

Books for further reading

BROWN, E. S. (1955). *Life in Fresh Water*. Oxford University Press, London

CARPENTER, K. E. (1928). *Life in Inland Waters*. Sidgwick & Jackson, London

CLEGG, J. (1965). *The Freshwater Life of the British Isles*. Warne, London

DOWDESWELL, W. H. (1959). *Practical Animal Ecology*. Methuen, London

ELTON, C. S. (1966). *Animal Ecology*. Methuen, London

ENGELHARDT, W. (1964). *Pond Life*. Burke, London

JANUS, H. (1965). *Molluscs*. Burke, London

MACAN, T. T. and WORTHINGTON, E. B. (1951). *Life in Lakes and Rivers*. Collins, London

MACAN, T. T. (1959). *A Guide to Freshwater Invertebrate Animals*. Longmans, London

MACAN, T. T. (1963). *Freshwater Ecology*. Longmans, London

MACFADYEN, A. (1963). *Animal Ecology*, Pitman, London

MELLANBY, H. (1963). *Animal Life in Fresh Water*. Methuen, London

MELLANBY, K. (1967). *Pesticides and Pollution*. Collins, London

ODUM, H. T. (1963). *Ecology*. Holt, Rinehart & Winston, New York

References

AMBÜHL, H. (1959). 'Die Bedeutung der Strömung als ökologischer Faktor.' *Schweiz. Z. Hydrol.*, **21**, 133–264

ANDREWARTHA, H. G. (1961). *Introduction to the Study of Animal Populations*. Methuen, London

BAILEY, N. T. J. (1959). 'Improvements in the interpretation of recapture data.' *J. Anim. Ecol.*, **21**, 120–7

BALFOUR-BROWNE, F. (1958). *British Water Beetles*, Vol. III, Ray Soc.

BISHOP, O. N. (1966). *Statistics for Biology*. Longmans, London

CARPENTER, K. E. (1926). 'The lead mine as an active agent in river pollution.' *Ann. appl. Biol.*, **13**, 395–401

CHAPMAN, D. G. (1954). 'The estimation of biological populations.' *Ann. Math. Stat.*, **25**, 1–15

DORIER, A. and VAILLANT, F. (1954). 'Observations et expériences relatives à la resistance au courant de divers invertebrés aquatiques.' *Trav. Lab. Hydrobiol. Grenoble*, **45** and **46**, 9–31

ELLIOTT, J. M. (1967). 'Invertebrate drift in a Dartmoor stream.' *Arch. Hydrobiol*, **63**, 202–37

FOX, H. M. and WINGFIELD, C. A. (1938). 'A portable apparatus for the determination of oxygen dissolved in a small quantity of water.' *J. Exp. Biol.*, **15**, 437–45

FRY, F. E. J. (1957). 'The aquatic respiration of fish.' *Physiology of Fishes*, Vol I, Academic Press Inc., New York

GREEN, J. (1961). *A biology of crustacea*. Witherby, London

GREENBERG, B. G. (1951). 'Why randomize?' *Biometrika*, **7**, 319–32

HOAGLAND, D. R. (1948). *Inorganic nutrition in plants*. Chronica Botanica

HOEL, P. G. (1943). 'The accuracy of sampling methods in ecology'. *Ann. Math. Stat.*, **14**, 289–300

JACKSON, C. H. N. (1939). 'The analysis of an animal population.' *J. Anim. Ecol.*, **8**, 238–46

LESLIE, P. H., CHITTY, D. and CHITTY, H. (1953). 'The estimation of population parameters. An example of the practical applications of the method.' *Biometrika*, **40**, 137–69

LINDEMANN, R. L. (1941). 'Seasonal food-cycle dynamics in a senescent lake'. *Amer. Midland Nat.*, **26**, No 3.

LINDEMANN, R. L. (1942). 'The trophic–dynamic aspect of ecology.' *Ecology*, **23**, 399–418

MACAN, T. T. (1958). 'Methods of sampling the bottom fauna in stony streams'. *Mitt. int. Ver. Limnol.*, **8**, 21

MACAN, T. T. and MAUDSLEY, R. (1966). 'The temperature of a moorland fishpond.' *Hydrob.*, XXVII, 1–22

MANN, K. H. (1956). 'A study of the oxygen consumption of five species of leech'. *J. Exp. Biol.*, **33**, 615–26

MORTIMER, C. H. (1956). 'The oxygen content of air-saturated freshwaters, and aids in calculating percentage saturation.' *Mitt. int. Ver. Limnol.* **6**, figure 1

NICHOLLS, D. G., SHEPHERD, D. and GARLAND, P. B. (1967). 'A continuous recording technique for the measurement of carbon dioxide, and its application to mitochondrial oxidation and decarboxylation reactions.' *Biochem. Journal*,

NICOL, E. A. T. (1935). 'The ecology of a salt-marsh'. *J. Mar. Biol. Ass. U.K.*, **20**, 203–61

ODUM, H. T. (1957). 'Trophic structure and productivity of Silver Springs, Florida.' *Ecol. Monogr.*, **27**, 55–112

PACAUD, A. (1949). 'Elevages combines mollusques–cladocères. Introduction a l'étude d'une biocoenose liminique.' *J. Rech.*, **9**, 1–16

PENNINGTON, W. (1941). 'The control of the numbers of freshwater phytoplankton by small invertebrate animals'. *J. Ecol.*, **29**, 204–11

PHILIPSON, G. N. (1968). 'Ecological pyramids: A field study at Malham Tarn.' *S.S.R.* **50**, No. 171, 262–78

REYNOLDSON, T. B. (1958). 'Triclads and lake typology in northern Britain—qualitative aspects.' *Verh. int. Ver. Limnol.*, **13**, 320–30

REYNOLDSON, T. B. (1961). 'Observations on the occurrence of *Asellus* (Isopoda, Crustacea) in some lakes of northern Britain. *Verh. int. Ver. Limnol.*, **14**, 988–94

REYNOLDSON, T. B. and YOUNG, J. O. (1963). *J. Anim. Ecol.*, **32**, 175–91

SCHLIEPER, C. (1958). 'Physiologie des Brackwassers.' *Binnengewässer*, **22**, 217–348

SLOBODKIN, L. B. (1960). 'Ecological relationships at the population level.' *Amer. Nat.*, **94**, 213–36

TEAL, J. M. (1957). 'Community metabolism in a temperature cold spring.' *Ecol. Monogr.*, **27**, 283–302

YOUNG, J. O., MORRIS, I. G. and REYNOLDSON, T. B. (1964). *Arch. Hydrobiol.*, **60**, 366–73

Index

Numerals: pages, roman; figures, *italics*; plates, **bold**

Aedes detritus, in brackish water, 44
Alisma plantago-aquatica, habitat, 58,
  3(b)
  ovipositing by *Trianodes bicolor* and
    Coenagrion puella, 64
  stems as a microhabitat, 57, 63
Alkalinity, seasonal variations, 5, *2*
Amphibious persicaria, see *Polygonum
    amphibium*
Ancylastrum fluviatile, attachment in
    swift current, 75, *31*, 79–80
  in flood conditions, 80
  respiration, 30–1
Ancylus lacustris, distribution on leaves
    of *Nymphaea alba* and *Polygonum
    amphibium*, 63
  respiration, 30–1
Anguilla anguilla, as predator of
    Gammarus pulex, 75
  migration, xiv
Anisopus terrestris, in sewage filters, 106
Anodonta cygnea, respiratory and feed-
    ing current, 35, *15*
Anopheles claviger, in brackish water, 44
Arrowhead, see *Sagittaria sagittifolia*
Asellus aquaticus, and calcium con-
    centration, 48
  as food for *Dendrocoelum lacteum*, 93
  ecological niche, 58
  in colony of *Plumatella repens*, 75
Asterionella formosa, and concentrations
    of silicates, 14
Azolla filiculoides, 12

Baetis sp. nymph, means of attach-
    ment in current, 71, *29(a)*
  rhodani, diurnal rhythm of activity,
    84
Berula erecta, and population, xiv
Biochemical oxygen demand
    (B.O.D.), 109
Biomass, calculation, 95
  pyramids of, 94–5, *41*

Bladderwort, see *Utricularia*
Buoyancy, effect on aquatic plants, 18
Brackish water, effect of tides on
    distribution of organisms, 43
  fauna, 43 *et seq.*
  invasion by freshwater species, 44
  species rarely found, 44

Caddis, see *Phryganea*, *Trianodes*,
    Hydropsyche angustipennis, *Rhyaco-
    phila dorsalis*
Calcium, see Mineral salts
Callitriche platycarpa, in brackish water,
    46
  reduction in woody tissue, 18, *7*
  shape of leaves, 18
Calorie, definition, 96
Calopsectra sp., detritus feeder, 99
  dives and energy loss, 101
Calorific value, 96, 99
Canadian pondweed, see *Elodea cana-
    densis*
Carbon dioxide, in fresh water, 4–5
  seasonal variation, 5, *2*
Chaoborus sp., feeding and food
    relationships, 33, *14(b)*
  respiration, 33
  ubiquity, 46
Chemical factors, 2–5
Chironomus sp. gills, 33, *14(c)*
  haemoglobin, 41
  in stream, 69
Chloeon dipterum nymph, oxygen and
    rate of gill movement, 35–6, *16*
Cladophora sp., and chironomid lar-
    vae, 69
  in stream, 68
Coenagrion puella, ovipositing on
    Nymphaea alba and *Alismas plan-
    tago-aquatica*, 64
Collecting, methods of, benthos, 86,
    38
  general, 62

Common eel, see *Anguilla anguilla*

Common reed, see *Phragmites communis*

Common rush, see *Juncus communis*

Common valve snail, see *Valvata piscinalis*

Community, and populations, xiv

Corixa, as a brackish species, 44

Corophium lacustre, range in brackish water, 45, *20(a)*

Coryneura sp. larva, inhabiting mud, 80

Cottus gobio, as predator of *Gammarus pulex*, 75

Cover/abundance scale, of vegetation, 59

Cranefly, see *Tipula* sp.

Crenobia alpina, food and distribution, 47, *22(a)*
  temperature as a factor in distribution, 48

Culex pipiens, as a brackish species, 44

Curled pondweed, see *Potamogeton crispus*

Current, and erosion, 10
  creation of microhabitats, 76, *32*
  influence on distribution, 77–80, Table 2, 10–11
  influence on aquatic plants, 18–20
  percentage of species caught at different speeds, *33*
  Pitot tube for measuring, 87, Appendix III, 117

Cyclops fimbriatus, ubiquity of, 42–3

Damselfly, see *Coenagrion puella*

Daphnia, energy balance, 96
  obtusa, food requirements, 46–7

Decomposers, and food relationships, 98–9, *42*

Dendrocoelum lacteum, use in Precipitin test, 93

Density, and increase in buoyancy, 18

Dispersal, establishment of breeding populations, 53

Dixa maculata, larva, example of hygropetrical fauna, 68, *27(a)*

Dragonfly, nymph, hawker, respiration and food capture, 36, *17(a)*
  damsel, respiration and food capture, 36–7, *17(b)*

Drift, invertebrate, 82–7

Duckweed, see *Lemna minor*

Dytiscus marginalis, adult, respiration and locomotion, 28–9, *12(a)*
  larva of *Hydrarachna* sp. as parasite, 64–5
  larva, effect of temperature on distribution, 51, *12(b)*
  predator of *Hydrophilus piceus*, 56
  respiration and locomotion, 29
  semisulcatus, larva, effect of temperature on distribution, 51

Eared pond snail, see *Limnaea auricularia*

Ecdyonurus, nymph, current and oxygen consumption, 52
  forcipula, nymph, diurnal rhythm, 84
  venosus, nymph, means of attachment, 71 *29(b)*

Ecological efficiency, general, 101–2

Ecology, definition, xiii

Ecosystem, definition, xv, 57

Eiseniella tetrahedra, inhabitant of mud, 80

Elodea canadensis, flowering, 21
  uptake of potassium, 16
  vegetative reproduction, 21, 24

Energy, balance, 95–6
  flow, 97–9, *42*, 100–1
  Law of Conservation, 101
  loss from aquatic ecosystems, 101
  transfer, 102

Enteromorpha intestinalis, in brackish water, 46

Entrophication, definition, 108

Epilimnion, 9

Epistylis, in sewage filters, 105, *43(a)*

Escherichia coli, as indicator of organic pollution, 109

Erpobdella octoculata, effect of current, 78–9
  effect of flooding, 80
  egg cocoon, *34(b)*
  general, 69

Euglena viridis, effect of light, 18

Euryhaline species, definition, 45

Exton stream, general, 68–75, **7(a–b)**, **8(b)**

Flatworms, see *Dendrocoelum lacteum* and *Polycelis nigra*

Floating pondweed, see *Potamogeton natans*
Flooding, effects of, 80–2
Fontinalis antipyretica, general, 12, 68
Food, and predation, 46–7
  current in *Simocephalus* sp., 33, *14*(a)
  general, 26
  relationships, 89 *et seq.*, *39*
  of *Chaoborus* sp., 33, *14*(b)
  variation with growth, 91
  webs, general, 89
Food chains, and energy transfer, 100–1
  balance, 91
  general, 88
  methods of tracing, 93–4
Freshwater, limpets, see *Ancylus lacustris* and *Ancylastrum fluviatile*
  shrimps, see *Gammarus pulex* and *Mysis relicta*

Gammarus pulex, distribution in tidal and estuarine waters, 46
  effect of current, 71–6
  in colonies of *Plumatella repens*, 75
Gasterosteus aculeatus, feeding, 26
Gerris, spp., general, 70
  locomotion and respiration, 27
Gills, general, 3
  variety and use in totally aquatic species, 33–40
Glossiphonia complanata, effect of current 77–8
  general, 69, *34*(a)
Gnat, see *Chaoborus* sp.
Great diving beetle, see *Dytiscus marginalis*
Great pond snail, see *Limnaea stagnalis*
Great ramshorn snail, see *Planorbis corneus*
Great silver water beetle, see *Hydrophilus piceus*
Gyrinus natator, general, 70
  legs and use of surface film, 27, 28, *11*(b)

Habitat, factors, 1
  terminology, xiv, 57
Hard water, 4
Hemiptera, surface dwellers, 26
Heptagenia interpunctata, diurnal rhythm, 84

Histogram, construction of, 61, *25*(c)
Hottonia palustris, and *Hydrophilus picens*, *9*, **5**(**a**), 54
  pollination, 21
Hydrarachna sp. adult, 64, *26*(a)
  larva, 64, *26*(b)
  as parasite, 64–5
Hydrobia jenkinsi, and flooding, 80
  as food for larva of *Hydrophilus piceus*, 55
  in food chain, 88
Hydrocharis morsus-ranae, turions, 24, *10*
Hydrofuge hairs, 26, 28, 29
Hydrogen-ion concentration, 4
Hydrometra stagnorum, use of surface film, 26, *11*(a)
Hydrophilus piceus adult, 53, **4**(**b**)
  egg cocoon, 54, **4**(**a**)
  in food chain, 88
  larva, general, 55, *23*(a)
    growth, *24*
  migration, 56
  pupa, general, 56, *23*(b)
Hydrophytes, 25
Hydropsyche angustipennis larva, current and distribution, 76
  current and food, 80
  larva and pupa, case-building and food, 71, *29*(c)
Hygrophytes, 25
Hygropetrical fauna, general, 68

Ilyocoris cimicoides, trophic relationships, 91
Invertebrate drift, as food of fish, 82
  daily fluctuation in a Dartmoor stream, 84, *37*
  density, 83–4, Table 3
  methods of sampling, 82, *36*(a–b)
Iris pseudacoris, in brackish water, 46

Jenkin's spire shell, see *Hydrobia jenkinsi*
Juncus communis, in brackish water, 46,

Ladder transect, construction of 59–61, *25*(c)
Leaves, shape in aquatic plants, 18
Leeches, see *Erpobdella octoculata* and *Glossiphonia complanata*
Lemna minor and productivity, 99

Light and photosynthesis, 9–10, 16–18
  intensity in a stream, 83 *et seq.*
Limnaea auricularia, restricted range
  of, 43
Limnaea glabra, as a soft-water
  species, 47
Limnaea pereger, ubiquity, 43
Limnaea stagnalis, ubiquity, 43
  use of surface film, 30–1, *13(a)*
Limnaea truncatula, as a brackish
  species, 44
Limnodrilus sp., detritus feeder, 99
Lincoln Index on marking/recapture
  method, 66
Limnophora sp. larva, 69, *27(b)*
Lithophilous organisms, and cur-
  rent, 76
  definition, 57, 68
  in Exton stream, 71
Lumbricidae, see *Eiseniella tetrahedra*
Lumbricillus lineatus, in sewage filters,
  105–6, *43(b)*
Lumbricus rubellus, in sewage filters,
  105–6

Marking and recapture, Lincoln
  Index, 66
  methods of, 66
Mayfly, see *Ecdyonurus* spp., *Chloeon
  dipterum*, *Baetis* sp. and *Rhithrogena semicolorata*
Microphage, description, 68
Microhabitat, example of, xv, 57
  in mud and silt, 80
  in a stream, 76
  in Tiverton Canal, 61, Table 1
Midge, see *Chironomus* spp., *Dixa*
  spp., *Tanypus* spp. and *Coryneura*
Miller's thumb, see *Cottus gobio*
Mineral salts, amount added in fish
  farms, 111
  and aquatic plants, 3
  and distribution of fauna, 47–8
  and *Utricularia* spp., 14
  calcium, 4
  and freshwater molluscs, 40
  carbonate and aquatic plants,
    13, 15
  potassium, effect of temperature on
    uptake by plants, 16, *6*
  silicates, and fluctuation in phyto-
    plankton, 14, *5*
Mixohaline conditions, see Brackish
  water

Molluscs, operculates, gills, 39
  calcium and, 40
  pulmonates, respiration, 30–1
Mosquito, see *Aedes detritus*, *Culex
  pipiens* and *Anopheles claviger*
Moth flies, see *Psychoda*
Mud, inhabitants, 40–1
  oxygen lack, 40
Mysis relicta, temperature and distribution, 48, *22(b)*, xiv

Nais sp. in colony of *Plumatella repens*,
  30(b)
Nemacheilus barbatula, general, 69,
  28(a)
Nepa cinerea, general, 70, *28(b)*
  larva of *Hydrarachna* as parasite,
    64–5
Nitella clavata, accumulation of calcium, 15
Notonecta glauca, a brackish water
  species, 44
  feeding, 26
  respiration, 29, 30
Nutrient salts, see Mineral salts
Nymphaea alba, distribution of *Ancylus
  lacustris*, 63–4
  ecological niche, 58
  ovipositing of *Trianodes bicolor* and
    Coenagrion puella, 64
  position of stomata, 21

Oligochaeta, in filter beds, 105
Orb-shell cockles, see *Sphaerium* spp.
Ostracods in colony of *Plumatella
  repens*, *30(a)*
Ostrea virginica, effect of salinity on
  distribution and development, 44
Oxygen, and photosynthesis, 10
  and plant respiration, 16
  effect of weed growth, 18
  evolution by aquatic plants, 12
  general, 2
  measurement of concentration, 10,
    114, 115, Appendix II(a) and (b)
  seasonal variation, 5, *2*
  table of solubility, 113, Appendix I
  water saturated with 2, *1*

Palaemonetes varians, in brackish water,
  45
Partially aquatic animals, general,
  28–31

Pasteur, Louis, and water-borne diseases, 104

Pea-shell cockles, see *Pisidium* spp.

Phosphorus-32 in tracing food chains, 93

Phototaxis in aquatic nymphs, 84

Phragmites communis, in brackish water, 46

Phryganea sp. feeding, locomotion and respiration, 38, *18(a)*

Physa sp., detritus feeder, 99

Physical factors, general, 5, *et seq.*

Phytoplankton, and biomass, 95
and silicates, 14, *5*
as primary producers, 89
in running water, 67

Pisidium spp., as food for *H. piceus*, 56
as primary consumer, 89
detritus feeder, 99
respiratory and feeding current, 39–40, *19(b)*

Pitot tube, modified for measurement of current, 87, Appendix II, 117–8

Planarians, calcium and determination of distribution, 47–8

Plankton, effect of flooding, 82
sampler, 82–3, *36(b)*

Planorbis carinatus, as a hard-water species, 47

Planorbis corneus, and *Daphnia obtusa* feeding experiment, 46–7
respiration, 31, *13(b)*

Plea leachii, trophic relationships, 91

Plumatella repens, colonization of concrete blocks, 75, *30*

Pollution, 103, *et seq.*
assessment of, 109–10
by DDT, 107
by detergents, 108
by factory waste, 107
by nitrates and phosphates, 108
E. coli as indicator, 109
Salmo trutta and, 108
stages of recovery in a stream, 107

Polycelis nigra, distribution, 48, *21*
in lead-polluted water, 107

Polygonum amphibium, and pollination, 21
distribution of *A. lacustris*, 63–4

Polyzoa, general, 75

Pond skater, see *Gerris* spp.

Pondweeds, as primary producers, 89

Potamogeton crispus, turions, 24, *10*

Potamogeton natans, pollination, 21, *9*

Potassium, see Mineral salts

Prawn, see *Palaemonetes varians*

Precipitin test, 93–4

Populations, general, xiv
sampling, 65, 66

Predator/prey relationships, general, 91
density, 92, *40*

Primary producers, 89–91
consumers, 89–91

Productivity, estimation of, 99–100

Profile, to make, 59–61, *25(b)*

Protech oxygen meter, Appendix II(*b*), 116

Protozoa, gaseous exchange, 31–2

Psychoda, in sewage filters, 106

Radioactive isotopes in tracing food chains, 94

Ranunculus aquatilis, leaves, 20, *7(c)*

Ranunculus fluitans, leaf reduction, 18, *7(b)*

Rat-tailed maggot, see *Tubifera* sp.

Respiration in *Chaoborus*, 33
Dytiscus marginalis, 28–9, *12(a–b)*
Gerris spp., 27
Gyrinus natator, 28
Notonecta glauca, 29–30
operculate molluscs, 39
Protozoa, 31–2
pulmonate molluscs, 30–1
Tubifex, 41
Velia spp., 27

Rhithrogena semicolorata, effect of current, 52
use of gills, 52

Rhyacophila dorsalis larva and pupa, attachment in stream, 71, *29(a)*

Riccia fluitans, general, 12

River crowfoot, see *Ranunculus fluitans*

Rivularia minutula, in food chain, 88

Rotifers, living within colony of *Plumatella repens*, 75

Sagittaria sagittifolia, habitat, 58
leaf shape, 20, *8*

Salinity, effect on distribution and development of *Ostrea virginica*, 44

Salmo salar, migration to fresh water, xiv

Salmo trutta, and pollution, 108
influence of temperature on growth, 110

Salmon, see *Salmo salar*
Salt concentration, general, xiv
Samplers, high-speed plankton
    sampler, 82–3, *36(b)*
    shovel sampler, 86–7
    Surber sampler, 86, *38*
    surface net for invertebrate drift,
    82, *36(a)*
Sampling benthos, 86, *38*
    populations, 65–6
Secondary consumers, 89–91
Sewage, filters, 105
    general, 104
Shovel sampler, see Samplers
Silicates, see Mineral salts
Simocephalus sp. feeding current, 33,
    14(a)
Simulium sp. larva and pupa, distri-
    bution in stream, 71
    effect of temperature, 51, 53
Sphaerium spp. respiratory and feeding
    current, 39–40, *19(b)*
Sphaeroma, a brackish water species,
    45, *20(b)*
Standing crop, 95
Stone loach, see *Nemacheilus barbatula*
Stonewort, see *Nitella clavata*
Stream, regions of, 67–8
Surber sampler, see Samplers
Surface film, exploration of, 26–8
Swan mussel, see *Anodonta cygnea*

Tanypus sp. larva, inhabiting mud, 80
Tardigrades, living in colony of
    Plumatella repens, 75
Temperature, and aquatic plants, 16
    and distribution of *Crenobia alpina*,
    48
    and distribution of *Dytiscus mar-
    ginalis*, 51
    and distribution of *Dytiscus semi-
    sulcatus*, 51
    and distribution of *Mysis relicta*, 48
    and growth of *Salmo trutta*, 110
    and oxygen content, 2, 16
    general, xiv, 5–9
    relationship between air and water,
    7–9, *3*, *4*
Tertiary consumers, 89–91
Theodoxus fluviatilis, effect of current,
    79
Thermocline, 9
Thigmotaxis, in aquatic nymphs,
    84

Three-spined stickleback, see *Gas-
    terosteus aculeatus*
Tides, and distribution of brackish
    water species, 43
Tipula sp. larva, inhabiting mud, 80
Tiverton Canal, *6(a)*, vegetation
    maps, 58–61, *25(a–b–c)*
Trianodes, larva, locomotion, 39,
    18(b)
    bicolor, ovipositing on *Nymphaea alba*
    and *Alisma plantago-aquatica*, 64
Trichoptera, in aquatic drift, 94
    in lead-polluted water, 107
Triclads, calcium concentration and
    distribution, 48, *21*
    in aquatic drift, 84
Triops (Apus) cancriformis, and compe-
    tition, 53
    in temporary pools, 43
Tubifera sp. larva, inhabiting mud, 80
Tubifex, respiratory current, 41
Turions, of *Elodea candensis*, 21
    of *Hydrocharis morsus-ranae*, 24, *10(b)*

Utricularia spp., and mineral salts, 14
    feeding on small organisms, 14–5
    neglecta, **1(a–b)**
Valvata piscinalis, gill, 39, *19(a)*
Vegetation maps, 58–61, **25(a–b–c)**
Velia spp. locomotion and respiration,
    27
    currens, in stream, 70
Vorticella, colonies on *Plumatella repens*,
    30, *30(c)*
    in sewage filters, 105

Wandering pond snail, see *Limnaea
    pereger*
Water bears, see tardigrades
Water-borne diseases, and rate of
    flow, 105
    and temperature, 104
    Louis Pasteur, 104
Water bugs, see Hempitera
Water crickets, see *Velia* spp.
Water crowfoot, see *Ranunculus aqua-
    tilis*
Water fern, see *Azolla filiculoides*
Water flea, see *Simocephalus* sp. and
    Daphnia obtusa
Water louse, see *Asellus aquaticus*
Water measurer, see *Hydrometra stag-
    norum*
Water mite, see *Hydrarachna* sp.

Water moss, see *Fontinalis* sp.
Water plantain, see *Alisma plantago-aquatica*
Water scorpion, see *Nepa cinerea*
Water soldier, see *Stratiotes aloides*
Water starwort, see *Callitriche platy-carpa*
Water violet, see *Hottonia palustris*
White water lily, see *Nymphaea alba*

Window fly, see *Anisopus fenestris*
Winkler method of determining oxygen, 10, Appendix II(*a*), 114–16

Yellow flag, see *Iris pseudacorus*

Zonation, study of communities, 59
Zones of a pond, 25
Zooplankton, and biomass, 95